MW01591700

Kinematics

William J. Patton

Reston Publishing Company, Inc.
A Prentice-Hall Company
Reston, Virginia

LIBRARY OF CONGRESS CATALOGING IN PUBLICATION DATA

PATTON, W. J.
   KINEMATICS.
   INCLUDES INDEX.
   1.   KINEMATICS.   I.   TITLE.
QA841.P37   531'.112'0151   78-31273
ISBN 0-8359-3693-7

© 1979 BY RESTON PUBLISHING COMPANY, INC.
A PRENTICE-HALL COMPANY
RESTON, VIRGINIA 22090

ALL RIGHTS RESERVED. NO PART OF THIS BOOK MAY BE
REPRODUCED IN ANY WAY, OR BY ANY MEANS, WITHOUT
PERMISSION IN WRITING FROM THE PUBLISHER.

10   9   8   7   6   5   4   3

PRINTED IN THE UNITED STATES OF AMERICA

Contents

Introduction

Every generation of mankind has been fascinated by motion: we speak of the "poetry of motion." Kinematics is the study and analysis of the functional motions of mechanisms, and because this subject is concerned with motion it is perhaps a more appealing subject than its sister-subjects, Statics and Dynamics.

This book presents the fundamentals of kinematic analysis. Though the presentation of kinematic principles is necessarily the conventional one, the scope of kinematics is suggested to the student by the inclusion of a number of examples taken from athletic activities, biomechanics, military applications, and automation. The final chapter extends the fundamental principles into those applications that involve automation devices; here the student will find considerable integration of simple statics, kinematics, and dynamics in the study of the more common automation devices. However, throughout this presentation no recourse is made to any advanced mathematics, not even calculus. As an aid to comprehension and to resolve the learner's doubts, sufficient worked examples are included.

In a first course in kinematics, some instructors prefer to omit a study of acceleration. This seems unfair to the student. As the text explains, accelerations are the critical element in mechanisms because they are the source and cause of many machine problems.

chapter one

Kinematics: Mechanisms

1.1. KINEMATIC ANALYSIS.

Kinematics is the science of motion: of displacement, velocity, and acceleration. Motion must be produced by forces, but the analysis of the forces that cause motion is a separate subject, called kinetics or dynamics. Newton's law relating force F to acceleration a through mass M, $F = Ma$, relates kinematics and dynamics.

Figs. 1-1 and 1-2 are two examples of machines requiring kinematic analysis. Fig. 1-2 shows the complex linkage, called a valve gear, that controls the movement of the valves admitting steam to the cylinders of a steam locomotive. The design of this valve gear was one of the most difficult kinematic problems to solve and was continually improved until the demise of the steam locomotive about 1950.

Kinematics has for its practical applications the movements of mechanisms and their design. Mechanisms are mechanical systems, that is, combinations of machine members that perform required motions in machines. A mechanism is required to open and close the valves of an engine, to operate the human arm and hand, to control the bucket of a front-end loader or backhoe, to cut crops with a swather, to retract and extend the landing gear of an aircraft, to position a rock drill to drill a hole at the proper angle, or to produce

FIG. 1-1   AN AUTOMATIC ASSEMBLY OPERATION BUILT
AROUND AN INDEXING TABLE. THE DESIGN
OF SUCH MACHINES REQUIRES THE COORDI-
NATION OF THE MOVEMENTS OF MANY
ASSEMBLY COMPONENTS AND WORKHEADS.

the many functions of a typewriter, a tape punch, or a card punch.
The members or *links* in such a mechanism are almost always rigid
bodies, that is, nondeformable. Though small deformations do occur
under the action of stress, they can almost always be ignored in
kinematic analysis.

Many parts-handling devices used in automatic assembly operations present fascinating problems in kinematic analysis. A number
of these devices will be examined later in this book. One of them is
the centerboard hopper of Fig. 7-2, and this may be used as an
example of the problems that arise in kinematic design.

A supply of components—rivets or pins for example—is
dumped into the hopper. In the center of the hopper is an oscillating
centerboard or pickup blade. The top surface of the blade is a track
shaped to fit the parts picked up when the blade rises through the
hopper. The blade catches a few parts as it rises, and when in its
highest position it is aligned with a delivery chute that guides the
parts to the assembly operation. The captured parts slide down the
track into the delivery chute.

The kinematic problem here is that of moving the blade
through the hopper to its highest position in the shortest possible

FIG. 1-2 ONE OF THE MOST COMPLEX KINEMATIC LIN-
KAGES EVER USED WAS THE MECHANISM,
CALLED A VALVE GEAR, THAT CONTROLLED
THE SLIDE VALVE ADMITTING STEAM TO THE
CYLINDERS OF A RAILROAD STEAM LOCOMO-
TIVE. THE VALVE IS DRIVEN BY THE CRANK
ON THE SECOND DRIVING WHEEL; THE
VALVE CYLINDER IS THE SMALLER CYLIN-
DER IMMEDIATELY ABOVE THE LARGER
STEAM CYLINDER.

time. To the nontechnical person this would appear to be a trivial
problem—simply operate the centerboard blade at the most furious
speed obtainable. But since the blade must be brought to a stop at
both its highest and its lowest position, a furious speed must call for
a furious acceleration and an equally sudden deceleration to the
highest position. However, too rapid a deceleration will simply cause
the parts to be tossed into the air by the blade. There is a definite
limit to the speed of operation, and as explained in Chapter 7,
finding this limit is not a trivial, nor even a simple problem in
kinematics. Such considerations suggest that the designer of this or
any other machine must proceed cautiously, anticipating problems
that can be solved only by a fundamental feeling for the principles of
kinematics. The design of the centerboard hopper is examined in Sec.
7.2, and will not be further discussed here.

In kinematic design, what is required is a certain displacement over a certain path. The speed or rate at which this displacement is accomplished is the velocity, and the velocity may have to be the highest velocity that circumstances permit or a velocity controlled to synchronize with the velocities of other machine members. To obtain a velocity, however, there must be an acceleration or deceleration, and, as intimated in the brief discussion of the centerboard hopper, accelerations may impose severe limitations on the scope of the design.

Accelerations are a matter of great concern because they are the cause of many problems. Velocities and displacements may be problems in themselves, but do not usually create additional problems. There is no problem in driving an automobile at 100 mph comparable to the problem of decelerating from that speed against a tree. To take a more serious example, consider the kinematic problem of a gun to be mounted in an army tank. The higher the muzzle velocity, the better the accuracy of the tank gun. Muzzle velocities as high as 3500 fps (feet per second) are in use, but 5000 fps would be welcome. But according to Newton's equation, $F = Ma$, a large acceleration requires a large force, and the tank shell would have to be accelerated to 5000 fps in an extremely small fraction of a second. To do this would require muzzle pressures of about 60,000 psi (pounds per square inch), and to contain such pressures the breech of the gun would be enormously heavy; if the gun is too heavy then the tank itself will be enormously heavy, and also any bridge strong enough to hold the tank while it crossed a river would be enormously heavy, and so the problems of acceleration expand. Quite often the kinematic designer wants a certain displacement and velocity, but doesn't much care what the acceleration is, however to design the parts of the machine it may be necessary to determine the accelerations in order to assess the forces that the machine parts must sustain.

Since the required motion of the mechanism will occur in two-dimensional space, or rarely, in three dimensions, graphical techniques are often convenient for representing displacement, velocity, and acceleration of parts of the mechanism. Graphical presentations are more easily visualized, understood, and checked than are mathematical analyses and often, too, are the means of avoiding complex and laborious computations, with their susceptibility to error. Graphical solutions will be relied upon rather heavily in the following pages because of convenience and ease of comprehension. But this is not to say that a graphical presentation is invariably more convenient or more acceptable. Quite often it is not. Where accuracy is required, graphics may be quite unsuitable. To obtain a higher degree of accuracy with graphical methods, lengths may be laid out with dividers, but even dividers are limited in the accuracy with which they are set. Often only a careful mathematical analysis is adequate.

Standards of accuracy should meet the needs of the problem. There are cases where a rough answer suffices. But even some familiar mechanisms, such as an engine piston and its connecting rod, require a surprising degree of accuracy. If the motion of the connecting rod is to move the piston to give a cylinder compression ratio of $1:10$, an error of about 1% in the displacement of the piston will produce a compression ratio of either $1:9$ or $1:11$, with a drastic effect on engine performance and efficiency.

The most outstanding achievements of the science of kinematics are undoubtedly the plotting of the trajectories for space flights such as the landing of men on the moon. Such work requires severe standards of accuracy. Perhaps we may conclude this introduction to the interesting subject of kinematics by reference to a kinematic claim that was actually an empty boast.

Early in World War II it was claimed that the B-17 Flying Fortress bomber, when equipped with the Norden bombsight, could drop a bomb in a pickle barrel from 30,000 ft. Let's examine this claim. A pickle barrel we will take to be 3 ft in diameter, therefore the maximum allowable bombing error would be half of this, or $1\frac{1}{2}$ ft. In order to hit the barrel, the position of the aircraft at the instant of bomb release would have to be known to a maximum error of $1\frac{1}{2}$ ft—by visual sighting, if you please. The bomber would be travelling at more than 250 fps. The timing for the bomb release, a manual operation, would have to be accurate within $1\frac{1}{2}/250$ sec or 6 msec (milliseconds). Here was a case where kinematic precision did not meet the needs of the problem. Further reference to this case is made in Chapter 2.

1.2.   THE FOUR-BAR LINKAGE MECHANISM.

The most famous, the most studied, and perhaps the most versatile of kinematic devices is the four-bar linkage of Fig. 1-3. This mechanism has a driver AB, a follower CD, and a coupler or connecting rod BC. The fourth member of the linkage is the base AD. Note that either AB or CD could be driver or follower. Several variations of the four-bar linkage are illustrated in Fig. 1-4. In the crank-rocker device, the shortest link is the driver, and it can rotate through a full circle while the follower swings through an arc. If the

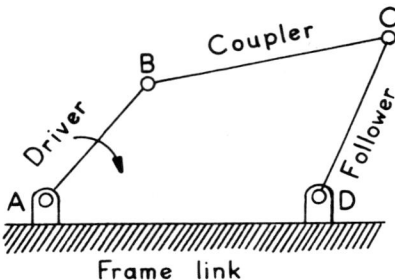

FIG. 1-3   A FOUR-BAR LINKAGE.

shortest link is the coupler, a double-rocker mechanism results. Finally, if the shortest link is the fixed link, then the resulting mechanism is a drag-link device.

As an aid to analyzing four-bar mechanisms, it is useful to consider the following relationships from Fig. 1-5:

$$AB\cos\alpha + BC\cos\beta + CD\cos\gamma = AD$$
$$AB\sin\alpha = BC\sin\beta + CD\sin\gamma$$

To introduce the four-bar linkage, consider the following example.

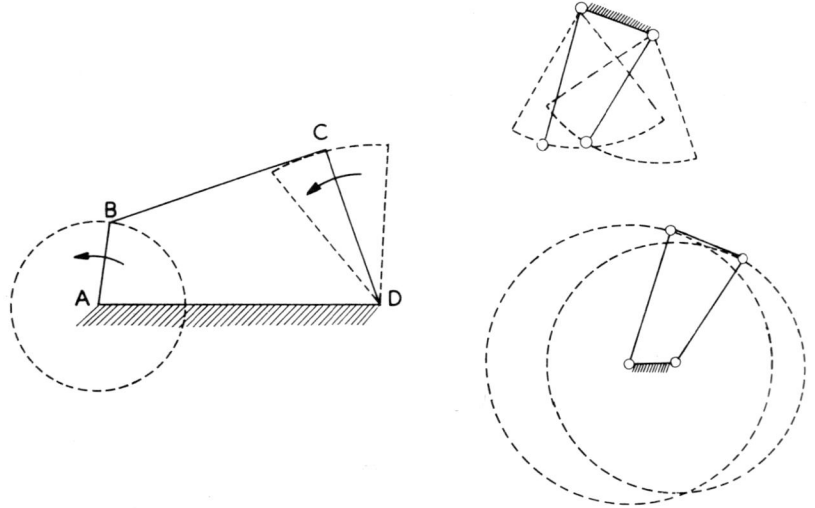

FIG. 1-4   VARIATIONS OF THE FOUR-BAR LINKAGE.
    (A)   CRANK-ROCKER MECHANISM.
    (B)   DRAG-LINK MECHANISM.
    (C)   DOUBLE-ROCKER MECHANISMS.

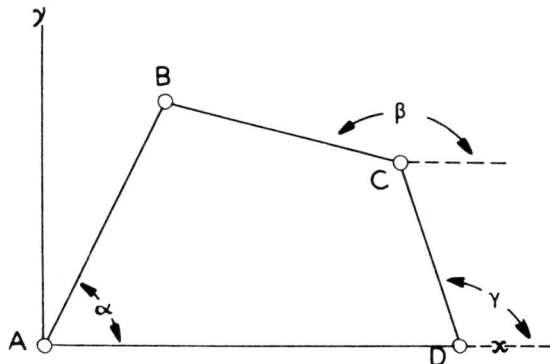

FIG. 1-5   GEOMETRY OF A FOUR-BAR LINKAGE.

Example.

Given:

$$AB = 4.4 \text{ cm}$$
$$CD = 10 \text{ cm}$$
$$BC = 16.2 \text{ cm}$$
$$AD = 15 \text{ cm}$$

Determine:

a)   The angle α between CD and AD when CD is in the extreme right position

b)   The angle β between CD and AD when CD is in the extreme left position

c)   The total angle of oscillation of CD for one revolution of AB.

Solution.

a)   For the extreme right position of CD, the links AB and BC must be extended on a straight line ABC. To find α, use the cosine law:

$$(AC)^2 = (AD)^2 + (CD)^2 - 2(AD)(CD)\cos\alpha$$
$$20.6^2 = 15^2 + 10^2 - 2(15)(10)\cos\alpha$$
$$\cos\alpha = -0.3312$$

Since $\cos\alpha$ is negative, α is larger than 90°. $\alpha = 109.3°$.

b)   For the extreme left position of CD, AB and BC must again lie on the same line, but AB overlaps BC as in Fig. 1-6.

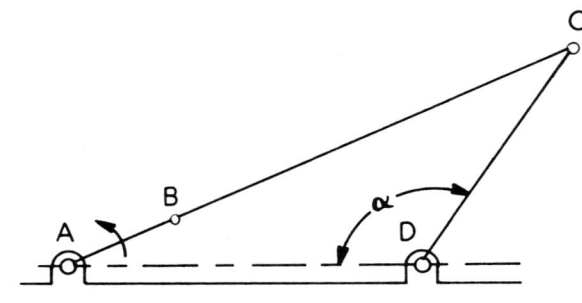

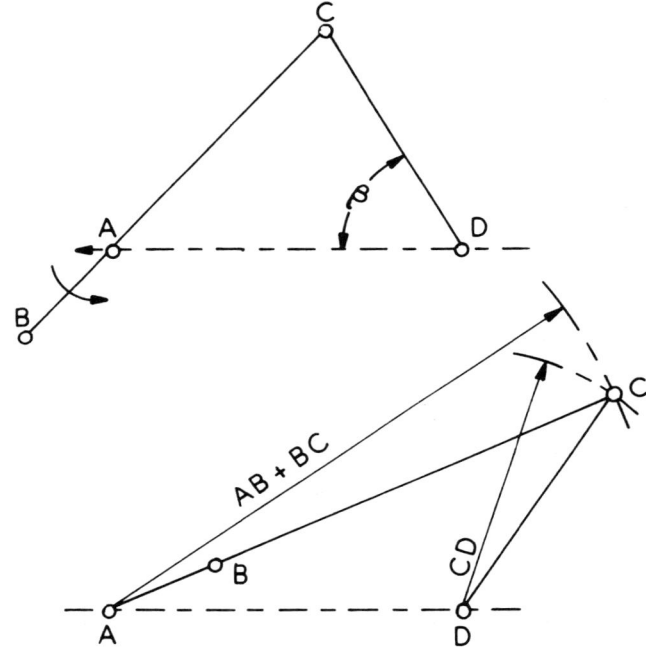

FIG. 1-6   ANALYSIS OF A FOUR-BAR LINKAGE.
  (A)   FOLLOWER *CD* IN EXTREME RIGHT-HAND
          POSITION.
  (B)   FOLLOWER *CD* IN EXTREME LEFT-HAND
          POSITION.
  (C)   GRAPHICAL CONSTRUCTION FOR LEFT-
          HAND POSITION OF *CD*. (SEE TEXT.)

$$(AC)^2 = (AD)^2 + (CD)^2 - 2(AD)(CD)\cos\beta$$

$$11.8^2 = 15^2 + 10^2 - 2(15)(10)\cos\beta$$

$$\cos\beta = 0.619 \quad \text{and} \quad \beta = 51.6°$$

c)   The total angle of oscillation of CD is $109.3 - 51.6 = 57.7°$.

Note that if AB rotates counterclockwise at a constant rotational speed, CD moves from angle β to α in a shorter interval of time than it requires to move from α to β. When CD moves from β to α AB moves through less than 180° of rotation. Four-bar linkages can be designed to give rapid movement in one direction and slow movement in the reverse direction.

This problem may also be solved graphically. Select a suitable scale factor. Draw an arc using CD as radius (see Fig. 1-6(c)). Then draw another arc using ABC as radius. The point of intersection of these two arcs gives the position of C when CD is in the extreme right-hand position.

1.3.   DESIGNING A FOUR-BAR LINKAGE

Designing a linkage to produce a required motion is more difficult than analyzing a linkage already designed. For an example of four-bar linkage design see Fig. 1-7. The driven link CD is to oscillate through an angle of 45° while AB rotates through one revolution. The length of CD is decided as 10 cm and the length AD may be any suitable length; 16 cm is selected. The lengths AB and BC must be determined.

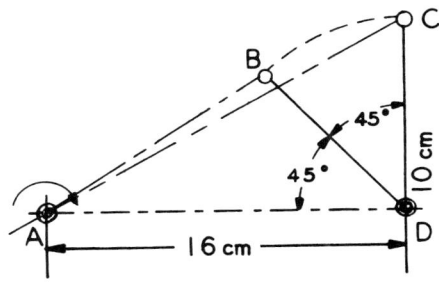

FIG. 1-7   DESIGN OF A FOUR-BAR LINKAGE TO ROTATE
FOLLOWER CD THROUGH 45°.

To solve such a design problem, use the cosine law for the two extreme positions of AC as before:

$$(AC)^2 = (CD)^2 + (AD)^2 - 2(CD)(AD)\cos 90°$$

But $\cos 90° = $ zero, and $AC = AB + BC$, so that

$$(AB + BC)^2 = 100 + 256 = 356$$
$$AB + BC = 18.87$$

Also
$$(AC)^2 = (CD)^2 + (AD)^2 - 2(CD)(AD)\cos 45°$$
where $AC = BC - AB$
$$(BC - AB)^2 = 10^2 + 16^2 - 2(10)(16)0.707$$
$$BC - AB = 11.39$$

There are two equations for AB and BC:

$$
\begin{array}{r}
AB + BC = 18.87 \\
-AB + BC = 11.39 \\
\hline
2BC = 30.26
\end{array}
$$

and
$$BC = 15.13 \text{ cm} \qquad AB = 3.74 \text{ cm}$$

In this example we have designed a four-bar linkage to supply rotary motion to a follower through $\frac{1}{8}$ of a revolution. By altering the lengths of the links, the follower could be made to rotate at the same speed as the driver, as shown in the parallel linkage of Fig. 1-8. A variant of this parallel four-bar linkage is given in Fig. 1-4(b), where, if one of the crank arms is driven at a constant speed, the other must rotate at a varying speed.

So far we have considered only the motion of the endpoints of crank, follower, and connecting rod. The versatility of the four-bar linkage is extended if we consider the motion of any point along the length of the connecting arm joining driver and follower. In Fig. 1-9 the motion of the midpoint between B and C is plotted. Since B has circular motion and C moves over a short arc, we would expect this

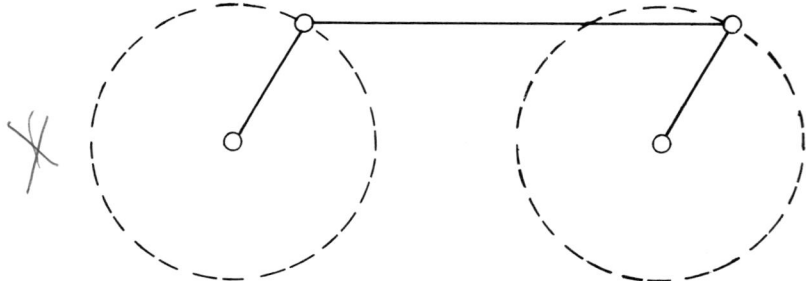

FIG. 1-8   FOUR-BAR LINKAGE WITH FOLLOWER ROTAT-
ING AT THE SAME SPEED OF THE DRIVER.

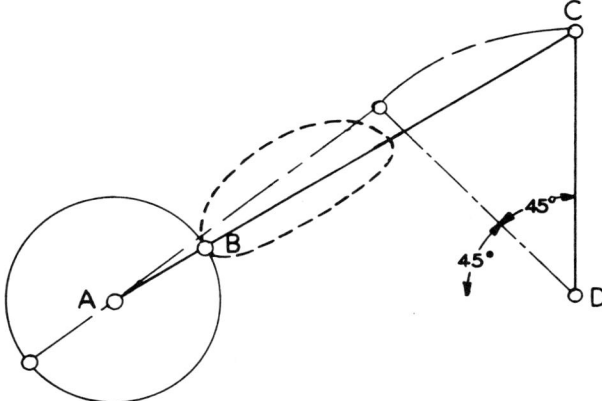

FIG. 1-9   PATH OF MIDPOINT BETWEEN *B* AND *C*.

midpoint to describe some curve that would be a compromise between the motions of the two endpoints. A portion of this curve of motion of the midpoint is nearly a straight line; possibly a true straight-line motion might be evolved, with a suitably designed four-bar linkage and a suitable selection of a point on the connecting rod.

Example.   An automatic work-handling device requires the work platform *BC* of Fig. 1-10 to be ~~translated~~ *MOVED* from position *BC* to position *B′C′*. Design a four-bar linkage to execute this movement.

Solution.   *BC* must be the connecting rod in the linkage. Fig. 1-10(b) shows the design method. Join *C′C* and *B′B*. Bisect both lines and erect a perpendicular bisector for both *C′C* and *B′B*. Arbitrarily choose points *A* and *D* on the two bisector lines as shown and draw a four-bar linkage *AB,BC,CD*. Scale off the link dimensions.

Points *A* and *D* may be located anywhere on the bisectors. If they are both located at the intersection *O* then *OB*, *BC*, and *OC* form a single link rather than a four-bar linkage.

The four-bar linkage is the basic link mechanism. It consists of a fixed member (the frame of the machine), a rotating driver and follower, and a connecting rod, or coupler. This mechanism offers a remarkable range of motions that may be obtained by adjusting the lengths of the four links. Different points on the connecting rod are capable of tracing an unlimited number of irregular curves. Virtually any path of motion can be obtained. The design process, however, is difficult; the four-bar linkage is a simple one only if you are not required to analyze it or to design it.

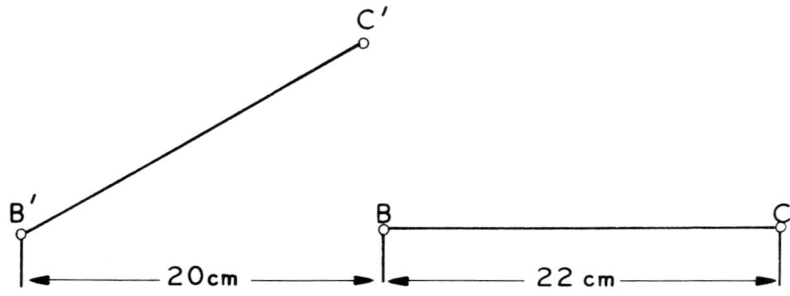

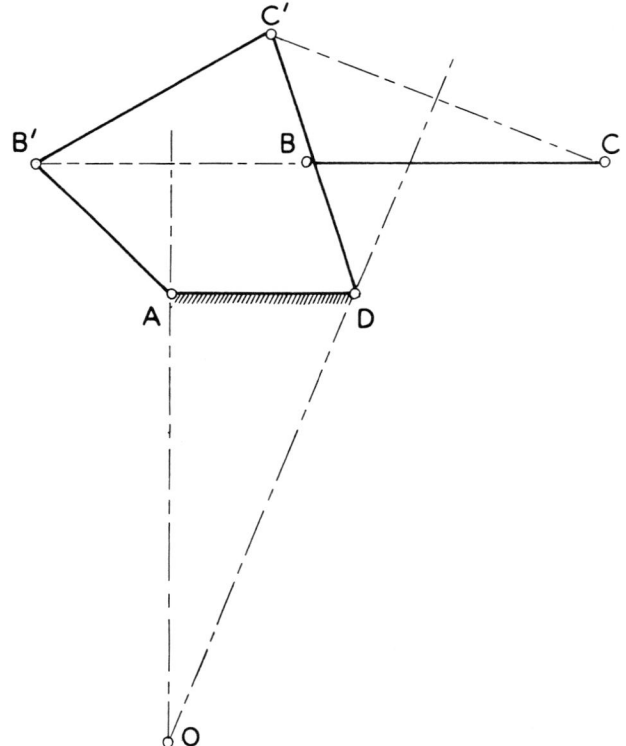

FIG. 1-10 (A) EXTREME POSITION OF A WORK PLAT-
FORM.
(B) THE REQUIRED FOUR-BAR MECHANISM.

1.4. THE SLIDER-CRANK MECHANISM.

The slider-crank mechanism of Fig. 1-11, familiar in internal combustion engines for connecting pistons to a revolving shaft, is a variant of the four-bar linkage.

The extended position of the linkage, that is, with the slider or piston in the extreme right position as in Fig. 1-11, is termed top dead center. The other extreme position of the slider or piston is termed bottom dead center. These terms are commonly abbreviated to TDC and BDC. From the figure it will be seen that the displacement or movement of the slider from TDC position is given by the expression:

$$d = (BC + AB) - AB\cos\theta - BC\cos\phi$$
$$= AB(1 - \cos\theta) + BC(1 - \cos\phi)$$

Another expression for d is commonly used; it is an approximation with a small error:

$$d = AB(1 - \cos\theta) + \frac{AB^2}{2BC}\sin^2\theta$$

The following example establishes the method of designing the movement of this linkage.

Example. (See Fig. 1-12.) The slider must reciprocate between two points, M and N, 15 cm apart. The shaft axis is placed on the line MN at a distance 22.5 cm from M as shown. Determine the lengths of both links.

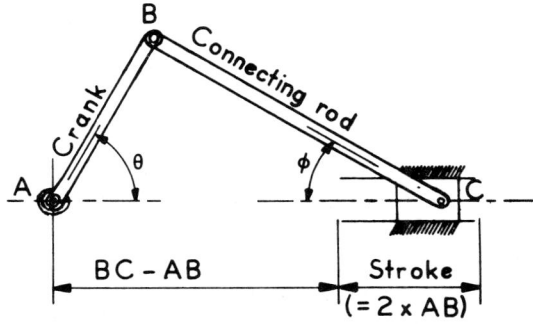

FIG. 1-11 SLIDER-CRANK MECHANISM.

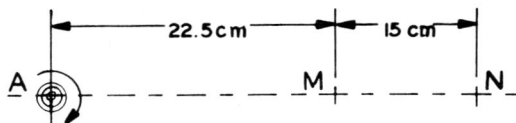

FIG. 1-12 DATA FOR DESIGN OF A SLIDER-CRANK. CRANKSHAFT AT A; M IS BDC AND N IS TDC.

Solution. To reach the extreme position N, the crank AB and the connecting rod BC must be extended in line. Therefore,

$$AB + BC = AM + MN$$
$$= 22.5 + 15 = 37.5 \text{ cm}$$

For the slider to reach the extreme position M, AB and BC will be overlapped:

$$BC - AB = AM = 22.5 \text{ cm}$$

Then
$$BC + AB = 37.5$$
$$BC - AB = 22.5$$
$$\overline{2BC = 60}$$

Therefore
$$BC = 30 \text{ cm} \quad \text{and} \quad AB = 7.5 \text{ cm}$$

1.5. THE SLIDING-LINK MECHANISM AND QUICK-RETURN MOTION.

The sliding-link mechanism is illustrated in Fig. 1-13. Its function is similar to that of a four-bar rocker. The crank L_1 rotates at a uniform angular velocity, and the slider on L_2 drives L_2 between its extreme positions A and B. For L_2 to oscillate between these two positions, L_1 must be shorter than the frame link; if the frame link is shorter than L_1, then L_2 will make complete revolutions (Fig. 1-14).

At the extreme positions A and B of Fig. 1-13 the crank L_1 is perpendicular to L_2. Therefore in moving L_2 from position A to position B, L_1 rotates through a large angle turning counterclockwise. To return L_2 to position A, L_1 rotates through a small angle. A mechanism with a shorter return time than its advance time is called a quick-return mechanism. When useful work is done on one stroke while the return stroke merely returns the mechanism to the initial or start position, a quick-return cycle is desirable. Examples are found in such machine tools as planers and shapers.

A complete sliding-block quick-return mechanism is shown in Fig. 1-15. Position B is BDC and position A is TDC. The angle ϕ is traversed by the crank on the advance stroke and angle θ on the return stroke, where $\phi + \theta = 360°$. Angle α is the total angular swing

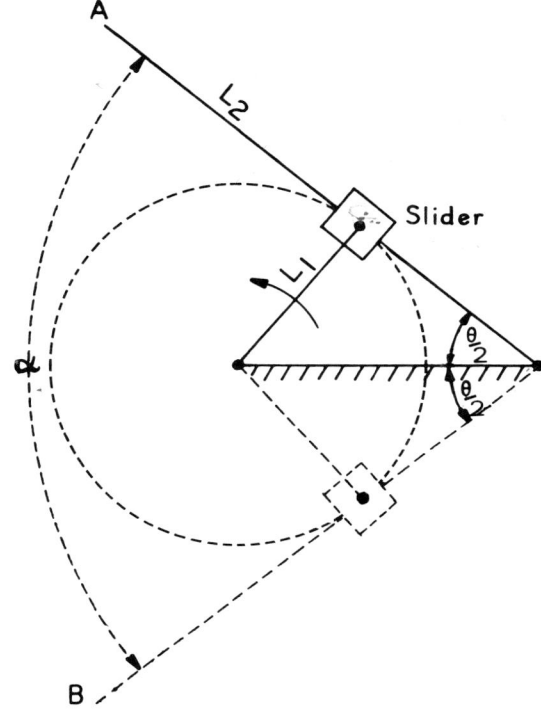

FIG. 1-13 THE SLIDING LINK MECHANISM.

of L_2. At the extreme positions A and B, L_1 and L_2 are perpendicular.

From the figure
$$\frac{\theta}{2} = 90° - \frac{\alpha}{2}$$
and
$$\theta = 180° - \alpha$$
where θ is the return stroke angle, usually less than 180°.

Also
$$\phi = 360° - \theta$$
$$= 360° - (180° - \alpha)$$
$$= 180° + \alpha$$
where ϕ is the advance stroke angle, usually greater than 180°.

To determine the ratio of speed of advance to speed of return, note that

$$\sin \frac{\alpha}{2} = \frac{L_1}{L_3}$$

$$\alpha = \frac{1}{2} \sin^{-1} \frac{L_1}{L_3}$$

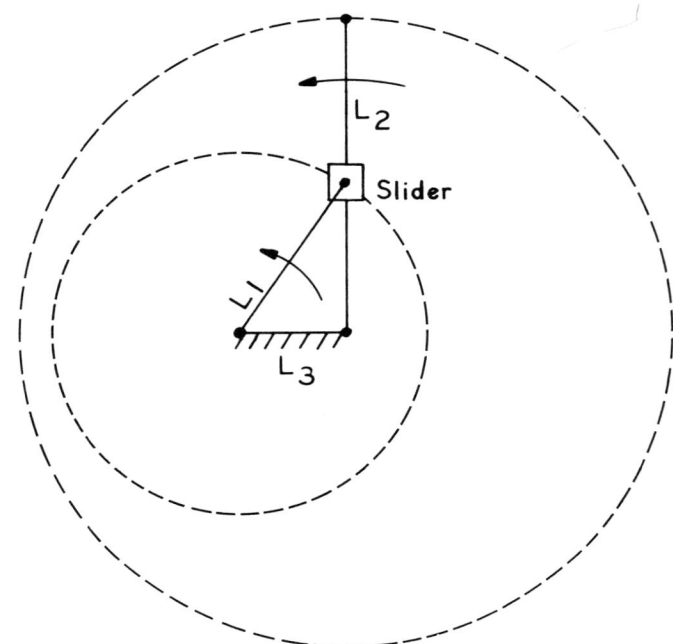

FIG. 1-14   SLIDING-LINK MECHANISM WITH GROUND
LINK L_3 SHORTER THAN CRANK L_1.

But since

$$\phi = 180° + \alpha$$

$$\phi = 180° + \frac{1}{2}\sin^{-1}\frac{L_1}{L_3}$$

and since

$$\theta = 180° - \alpha$$

$$\theta = 180° - \frac{1}{2}\sin^{-1}\frac{L_1}{L_3}$$

The ratio of these angles ϕ/θ is the ratio of speed of advance to speed of return:

$$\frac{\phi}{\theta} = \frac{180° + \frac{1}{2}\sin^{-1}L_1/L_3}{180° - \frac{1}{2}\sin^{-1}L_1/L_3}$$

The required stroke S is always known. In determining a relationship between length of links and stroke, we make the assumption that the ground link L_3 is perpendicular to the direction of movement of the load, as in Fig. 1-15. To assume otherwise makes for a more complex analysis, which is normally unwarranted. The stroke distance S is a chord between A and B subtending the arc AB

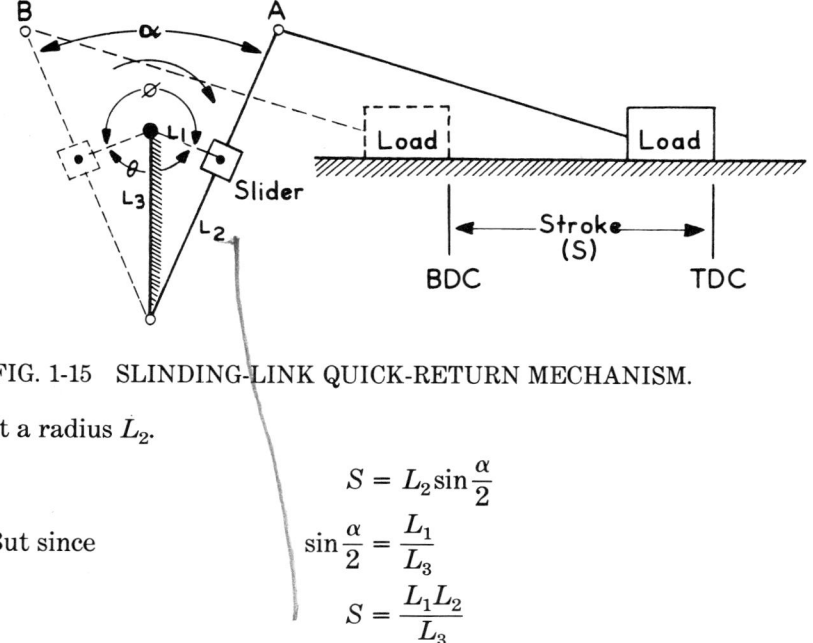

FIG. 1-15 SLINDING-LINK QUICK-RETURN MECHANISM.

at a radius L_2.

$$S = L_2 \sin \frac{\alpha}{2}$$

But since

$$\sin \frac{\alpha}{2} = \frac{L_1}{L_3}$$

$$S = \frac{L_1 L_2}{L_3}$$

1.6. STRAIGHT-LINE MECHANISMS.

Two methods for guiding a body or a point along a straight line are usual in machinery. The part that requires straight-line motion may be guided by an accurate plane surface as used in the bed of a lathe, a shaper, or a woodcutting table saw. An alternate method, though not as accurate, is the use of a pneumatic or hydraulic cylinder to push or pull the part in a straight line. Yet, there are many circumstances where these simple and effective methods cannot be employed, especially in instrumentation and automatic devices. The following are only a sampling of the many straight-line mechanisms available.

James Watt, the builder of steam engines, required a straight-line motion and devised the one shown in Fig. 1-16. It uses two cranks AB and CD, equal in length, each rotating about its end. These are pinned to the connecting rod BC. The midpoint of BC, designated E, displaces over a virtually straight path for some distance. Its complete path is shown.

A more accurate straight-line device is the isosceles straight-line linkage of Fig. 1-17. The lengths AB, BC, and BD are equal.

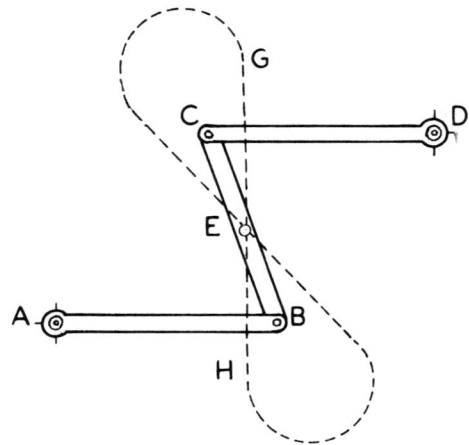

FIG. 1-16 WATT'S STRAIGHT-LINE LINKAGE.

Point *D* is guided along a straight line as shown. This line is perfectly straight throughout its range of movement.

The epicyclic straight-line mechanism of Fig. 1-18 uses a planet roller (or gear) *P* with a diameter equal to the radius of the stationary ring *R*. If *P*, mounted on the link, rolls on *R* without slipping, point *A* follows the straight line shown, which is a diameter of *R*.

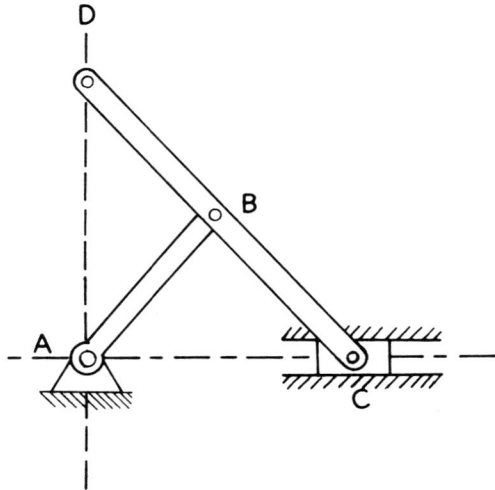

FIG. 1-17 ISOSCELES STRAIGHT-LINE LINKAGE.

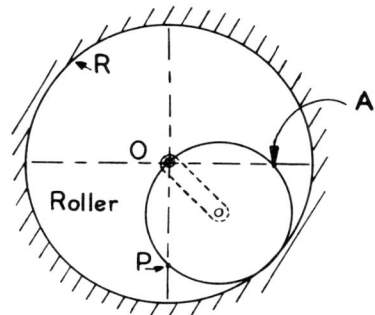

FIG. 1-18　EPICYCLIC STRAIGHT-LINE MECHANISM. THE
PIVOT FOR ROLLER *P* LIES ON THE LINK,
WHICH IS PIVOTED AT *O*.

PROBLEMS

1. In Fig. P1-1, the link *LM* is 20 cm long and point *P* is located 6.5 cm from *L*. The guided movement of *L* is perpendicular to the movement of *M*. Plot successive positions of the point *P* and connect them with a smooth curve.

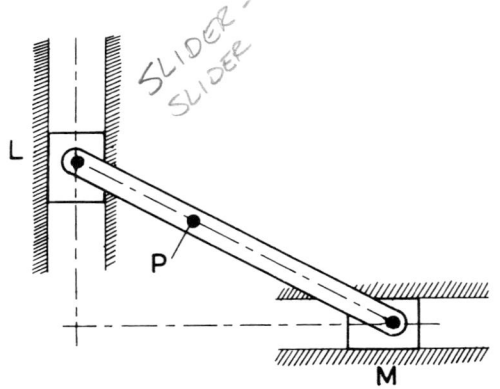

FIG. P1-1

2. Fig. P1-2 shows Tchebichev's approximate straight-line mechanism. *A* and *D* are fixed axes of rotation; *P* is the midpoint of the link *CB*. Dimensions are as follows:

    $$AD = 20 \text{ cm}$$
    $$AB = CD = 25 \text{ cm}$$
    $$CB = 17.3 \text{ cm}$$

    Draw Tchebichev's mechanism to half-size or larger and plot the path of *P* to find the extent of its straight-line motion.

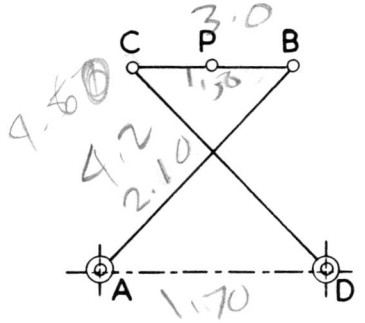

FIG. P1-2   TCHEBICHEV'S STRAIGHT-LINE MECHANISM.

3. In the linkage of Fig. P1-3, *A* and *B* are fixed axes of rotation. Dimensions are as follows:

$$AB = BC = 7.5 \text{ cm}$$
$$AE = AD = 10 \text{ cm}$$
$$DC = CE = EF = FD = 12.5 \text{ cm}$$

Draw the mechanism full size and plot the path of *F*.

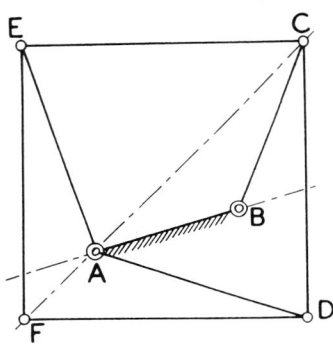

FIG. P1-3

4. A four-bar linkage with its dimensions is given in Fig. P1-4. Determine the angle that *CD* sweeps through in one revolution of *AB*.

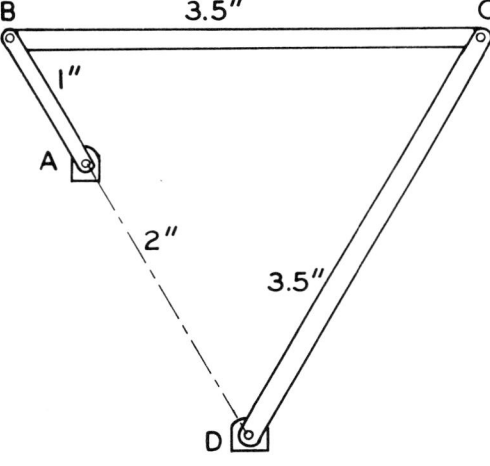

FIG. P1-4

5. In the mechanical summing mechanism of Fig. P1-5, a quantity X is set on the left-hand scale and a quantity Y is set on the right-hand scale. Explain why the middle scale will give the sum of X and Y.

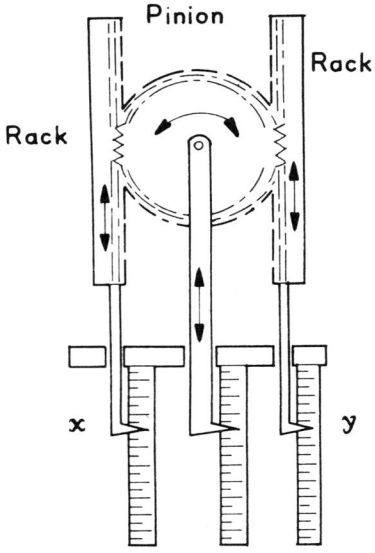

FIG. P1-5 MECHANICAL SUMMING MECHANISM.

6. A slider-crank mechanism in a small internal combustion engine has a crank length of 22 mm and a connecting rod with a length of 80 mm. The slider motion is in the same plane as the axis of rotation of the crank. What is the stroke of the piston?

7. The slider in a slider-crank mechanism has a stroke of 10 cm. The distance from the axis of the crank to BDC is 20 cm. Slider motion and the axis of the crank are in the same plane. Determine the lengths of connecting rod and crank
    a) Graphically
    b) Mathematically

8. When the crank of the slider-crank of question 7 has rotated 45 degrees off TDC, what is the displacement of the slider from TDC?

9. A four-bar linkage has the following dimensions:

Ground link	7.5 cm
Crank	2.5 cm
Connecting rod	5.0 cm
Follower	3.75 cm

Determine the extreme positions of the follower.

10. a) Determine the ratio of speed of advance to speed of return in the sliding-link mechanism shown in Fig. P1-10.
    b) Determine the stroke.

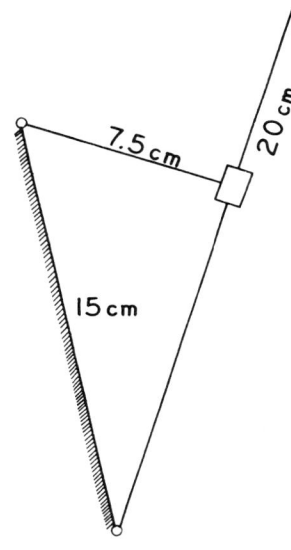

FIG. P1-10

11. In Fig. P1-11 crank *AB* rotates at a constant speed of 120 rpm. Determine the time required for *CD* to move from the extreme left to the extreme right position.

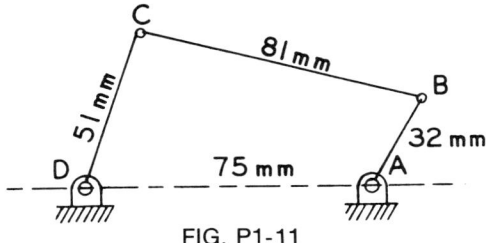

FIG. P1-11

12. The four-bar linkage of Fig. P1-12 is driven by the wheel. *BEC* is a single link. Dimensions are as follows:

wheel diameter 100 mm   *AB*=75 mm
     BC=83 mm   *CD*=37 mm
     DA=75 mm   *CE*=25 mm
     EF=50 mm   *A*=75 mm

Determine the extreme positions of *AB*.

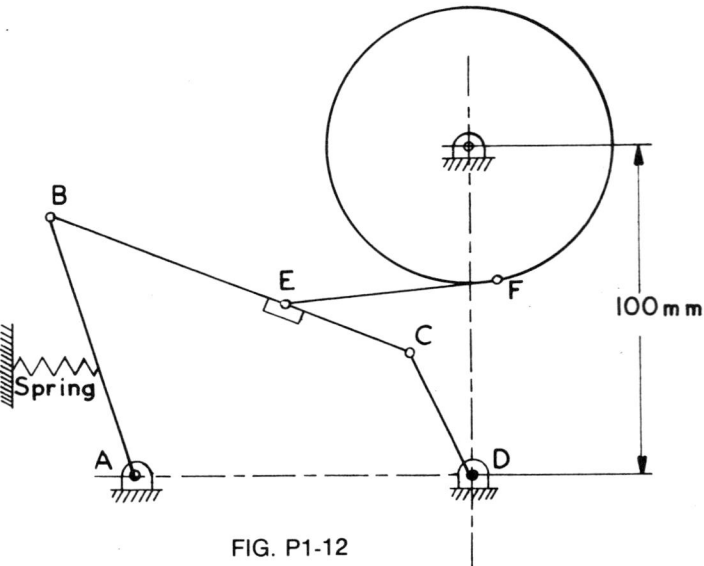

FIG. P1-12

13. Using graphical methods design a four-bar crank-rocker mechanism to oscillate the rocker through a range of 60° (see Fig. P1-13). Make the length of the rocker 40 mm long and the length of the ground link 110 mm (distance between rotational axes of crank and rocker).

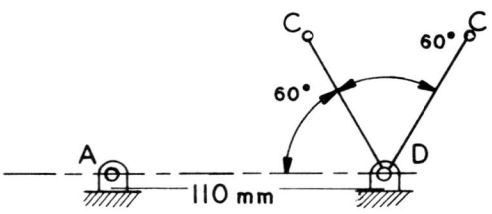

14. Solve Problem 13 mathematically.

15. A crank-rocker mechanism must be designed to index a 10-tooth ratchet with a diameter of 30 mm. The ratchet advances one tooth with each revolution of the crank *AB*. Make *CD* 40 mm long and *AD* 90 mm long. Determine the lengths of the crank and the connecting rod.

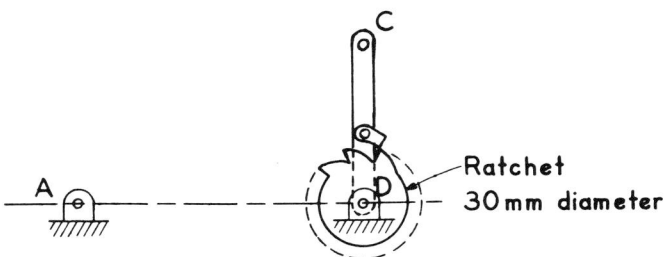

16. Fig. P1-16 shows basic dimensions of a crank-rocker mechanism to be designed. The rocker *AB* must oscillate through 70°, as shown, when the crank *DC* rotates one revolution. Graphically determine the lengths *CB* and *CD*.

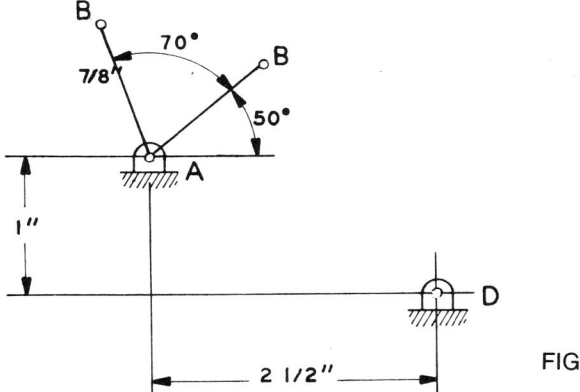

FIG. P1-16

17. Solve Problem 16 mathematically.

18. Design the four-bar linkage shown in Fig. P1-18 to oscillate the follower *CD* over the angular range shown.

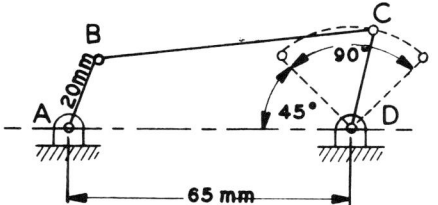

FIG. P1-18

19. Design a four-bar linkage to move the platform *BC* between the extreme positions shown in Fig. P1-19.

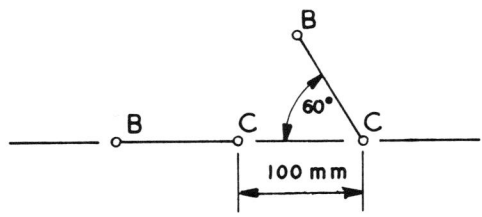

FIG. P1-19

20. Dimension a sliding-link mechanism (Fig. 1-13) with a ratio of advance time to return time of 2:1 and a stroke of 300 mm. Link L_1 in Fig. 1-13 is to be 100 mm long and rotates at a constant angular velocity. Note that the required ratio of 2:1 requires that the return stroke be executed in 120° of revolution of L_1. Why?

21. Design a linkage to drive a slider between the points *B* and *T* in a straight line as shown in the figure. The linkage is to be driven by a shaft with axis at *A* and continuously revolving in the same direction.

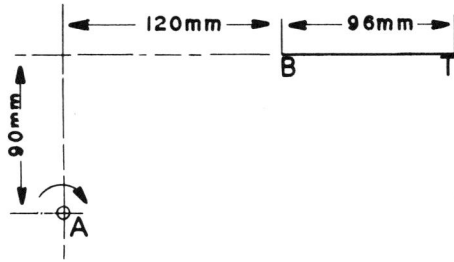

FIG. P1-21   RECIPROCATING MECHANISM.

22. Design a mechanism to oscillate the crank *CD* in Fig. P1-22 through an angle of 90°. The length of *CD* is to be 90 mm. The mechanism is to be driven by a shaft located at *P* and continuously revolving.

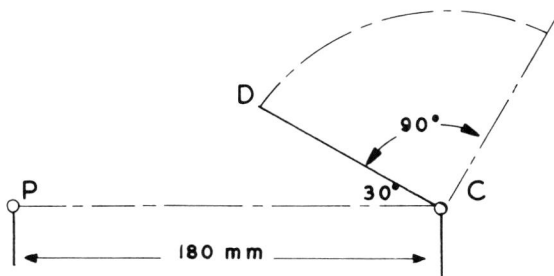

FIG. P1-22 OSCILLATING MECHANISM.

23. Design a linkage to drive an oscillating arm 150 mm long, the range of oscillation to be 75 degrees. The time ratio of the forward to the return stroke is to be 2:1. A revolving shaft turning at a constant speed is to drive the linkage.

chapter two

Basic Motion Analysis

2.1. SCALAR AND VECTOR QUANTITIES.

A quantity whose only characteristic is magnitude is called a scalar quantity. Examples of scalar quantities are temperature, time, area, volume, money, and horsepower. Scalar quantities are manipulated by the usual methods of arithmetic: two areas can, for example, be added directly to find their scalar sum. The scalar quantity in addition to magnitude may have sense, that is, may be positive or negative. Vector quantities have also both magnitude and sense but are somewhat more complex.

A vector quantity or *vector* is a quantity with the characteristics of magnitude, direction, and sense (positive or negative). Force, stress, strain, displacement, velocity, and acceleration are all vector quantities. Consider the case of Fig. 2-1, where a 100-lb weight is suspended by a 100-lb pull in each of two cables at 30° to the horizontal. If the two cable forces were scalar quantities, then the sum of the two would be 200 lb if both were taken as positive or zero if one is positive and one is negative. Either scalar total is wrong because it ignores an essential characteristic of a vector, its direction. The two forces must be summed by incorporating direction in the manipulation, as in Fig. 2-1. Actually they sum to 100 lb resultant force. Mathematically the sum of the two force vectors can be represented as $A \mapsto B = C$. The vector plus symbol indicates vector addition of A and B preserving direction. Note also that

$$A \mapsto B = B \mapsto A$$

Vectors may be represented by a line drawn to any convenient scale to represent the scalar quantity of the vector and at the angle of the vector. The sense of the vector is represented by putting an arrowhead at the proper end of the line.

The method of vector addition is as follows. Suppose that vector B must be added to vector A.

$$A \mapsto B = C$$

The vector addition symbol indicates that A and B, and also C, are vector quantities, and therefore the directions of A and B must be preserved. Addition is made by placing the tail of vector B at the head of vector A. The vector C is the length from the origin of A to the terminal end of B as shown. The numerical value of C may be obtained by trigonometry or by scaling the length.

If B must be subtracted from C (see Fig. 2-2),

$$C \mapsto (-B) = C \rightarrow B$$

The vector $-B$ is 180° out of phase with the vector $+B$.

Example.   Two vectors A and B are shown in Fig. 2-3. Completely determine the vector C equal to the sum of A and B.

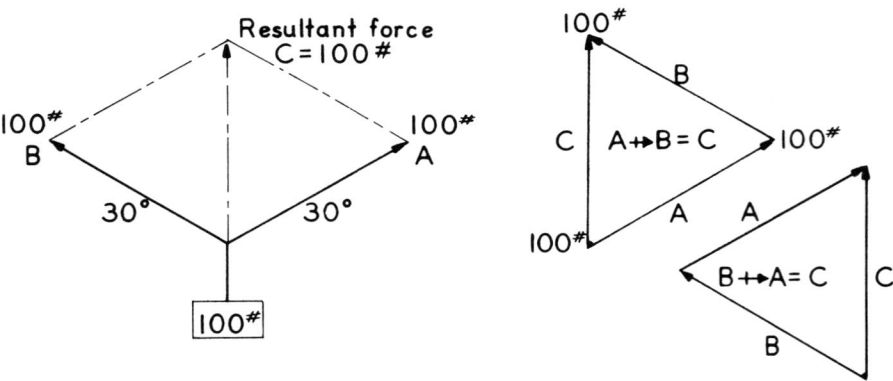

FIG. 2-1   FORCE VECTORS AND VECTOR TRIANGLES.

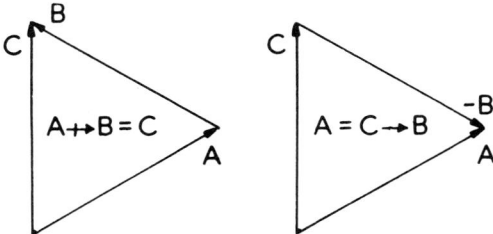

FIG. 2-2   VECTOR SUBTRACTION.

Solution.   The vector triangle is shown in Fig. 2-3. B is added
to A by attaching the tail of B to the arrow of A. The resultant
vector C is the vector from the initial point of A to the terminal
point of B.

To determine C use the cosine law:

$$(OQ)^2 = (OP)^2 + (PQ)^2 - 2(OP)(PQ)\cos(OPQ)$$

But angle $$OPQ = 90° - 25° = 65°$$

$$(OQ)^2 = 10^2 + 7^2 - 2(10)(7)0.423$$

and $$OQ = 9.48.$$

To find θ use the sine law:

$$\frac{OQ}{\sin(OPQ)} = \frac{OP}{\sin(PQO)}$$

$$\frac{9.48}{\sin 65°} = \frac{10}{\sin(PQO)}$$

$$\sin PQO = 0.957 \qquad \text{and} \qquad \text{angle } PQO = 73°$$

Therefore $$\theta = 90 - 73 = 17°$$

Any number of vectors may be added together. Thus in Fig.
2-4, $A \mapsto B \mapsto C = D$. Such a vector configuration is called a vector
polygon.

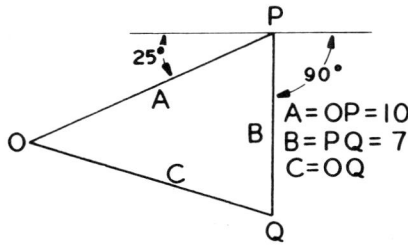

FIG. 2-3   VECTOR TRIANGLE TO BE SOLVED FOR VEC-
        TOR C.

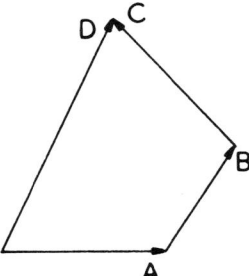

FIG. 2-4 VECTOR POLYGON.

Sometimes the X and Y components of a vector are required, as in Fig. 2-5. Frequently the resolution of a vector at some angle into its components parallel to a pair of axes simplifies complex trigonometric calculations.

Vector quantities may be multiplied or divided by a scalar quantity. The result is still a vector quantity, since only a change of scale is involved.

Parallel vectors may be added as scalar quantities. Thus if an aircraft is flying due north at an air speed of 130 mph against a north wind of 30 mph, its speed relative to the ground is $130-30$ or 100 mph.

2.2. ABSOLUTE AND RELATIVE MOTION.

The surface of the earth, or an arbitrary point on its surface, is normally chosen as a fixed reference for motion. In graphical work either a given point or a pair of axes are the datums, or references,

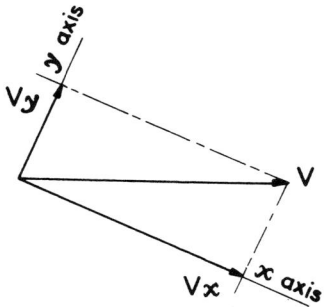

FIG. 2-5 X AND Y COMPONENTS OF A VECTOR V.

for points in motion. Motion relative to such fixed references is called *absolute motion*. Motion relative to another body in motion is called *relative motion*.

Consider the case of an aircraft about to make a landing on the flight deck of an aircraft carrier. The deck of the carrier may be the fixed reference for motion of the aircraft, or the surface of the earth may be that reference. In the latter case, the absolute motion of the aircraft—its motion relative to the earth considered as fixed—has two components: the velocity of the aircraft relative to the carrier and the velocity of the carrier relative to the earth's surface. That is

$$v_A = v_C \mapsto v_{a/c}$$

where $v_{A/c}$ = velocity of the aircraft relative to the carrier
v_C = velocity of the carrier relative to the earth
v_A = velocity of the aircraft relative to the earth

Supplying some numerical values, assume

$$v_C = 30 \text{ mph due north}$$
$$v_{A/c} = 100 \text{ mph also due north}$$

Then $v_A = 130 \text{ mph}$

If the carrier were motionless, then $v_C = 0$ and $v_{A/c} = v_A$.

A navy man would of course remind us that naval kinematics measures speed in knots, not miles per hour.

We may draw upon a more familiar situation to clarify the concept of relative velocity. Suppose automobile B follows automobile A down the freeway. If both A and B have the same velocity, then their relative velocities are both zero:

$$v_{A/B} = v_{B/A} = 0$$

To an observer in car A, car B does not change position with respect to A, and therefore the relative velocity of B with respect to A is zero. The same holds true for the inverse case of A with respect to B.

Next, assume that the leading vehicle A is travelling 60 mph and the trailing vehicle B 40 mph. The relative velocity is the velocity of the other vehicle as viewed from one vehicle. In this case if the observer is in B, then car A is travelling 20 mph faster:

$$v_{A/B} = 20 \text{ mph}$$

But if the observer is in car A and watching car B, then car B is falling behind (moving away from car A) at 20 mph, and

$$v_{B/A} = -20 \text{ mph}$$

Then $v_A = v_B \mapsto v_{A/B}$
$$60 = 40 + 20$$

Conversely,

$$v_B = v_A \mapsto v_{B/A}$$
$$40 = 60 - 20$$

since
$$v_{B/A} = -v_{A/B}$$

A peculiar impression of relative velocity occurred in the days of railroad travel. If a passenger was sitting in a train about to move and looked out of his window at the train standing on the next track, as his train began to move he would have no sensation of motion. Instead he would invariably see the other train as moving backward. He sensed his own absolute velocity as the relative velocity of the other train to his own train.

The special case of relative velocity of two points on a rigid body must be noted. A rigid body is understood to be a body that does not stretch, contract, or deform. The small deformations of bodies due to stress can be ignored in kinematic analysis, with the obvious exceptions of springs and rubber components, which are designed to deform. Consider the rigid body AB of Fig. 2-6. If B has a displacement, velocity, or acceleration with respect to A, there can be no component of these in the direction AB, for then the body would not be rigid. If there is a displacement, velocity, or acceleration of A with respect to B or B with respect to A, then it must be at right angles to a line joining A and B. A can rotate about B, or B about A, but cannot translate with respect to the other point.

Example. Two lift trucks A and B in a warehouse both move 10 ft as shown in the figure. What is the relative displacement (that is, movement) of B with respect to A?

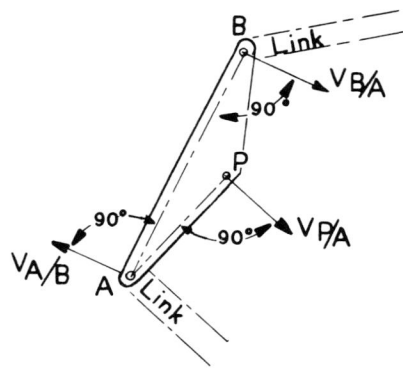

FIG. 2-6 RELATIVE VELOCITIES IN A RIGID BODY.

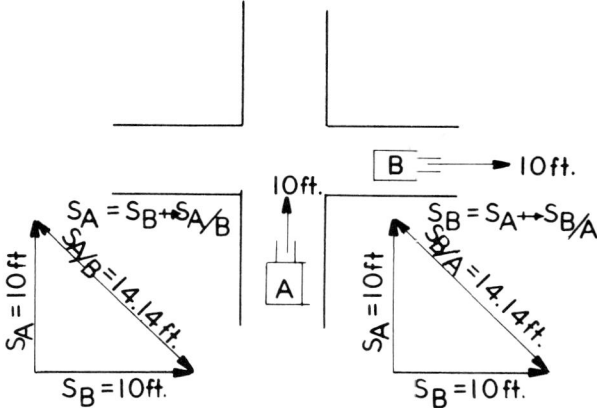

FIG. 2-7 RELATIVE DISPLACEMENT OF TWO WAREHOUSE LIFT TRUCKS.

Solution. $S_B = S_A \mapsto S_{B/A}$ See the vector diagram, Fig. 2-7.

The absolute displacement of B (displacement of B relative to the earth) is to the right, but to an observer on truck A, B seems to be moving backward at an angle because A is moving forward.

Note in the vector diagram that the absolute displacements originate from the origin of the vector diagram.

2.3. DISPLACEMENT.

The displacement of a particle or body is a vector quantity that measures a change in position as a result of motion. This displacement is independent of the path taken by the body from the original to the final position. Thus, Fig. 2-8 shows three different paths for a displacement from P_1 to P_2, but for all three cases the displacement is the vector shown, P_1P_2.

Displacement is usually represented by the symbol s. Units are length units: cm, km, ft, mi, etc.

The scalar quantity called distance measures the length of the path travelled. Distance, as in Fig. 2-8, may be a much larger quantity than displacement.

Consider now, point P moving along a path from P_1 to P_2 as in Fig. 2-9. Point O is any point arbitrarily chosen. The distances of P_1 and P_2 from O are represented by position vectors s_1 and s_2. The

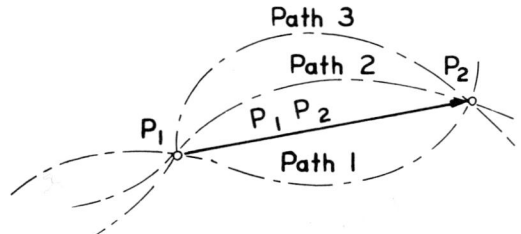

FIG. 2-8   DISTANCES AND DISPLACEMENT VECTOR.

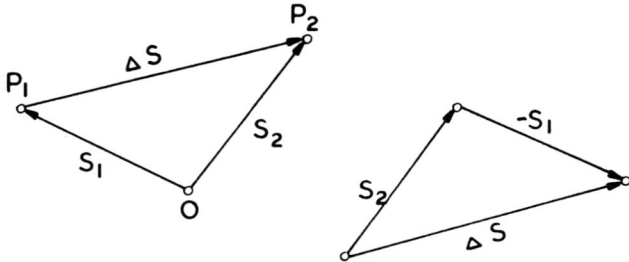

FIG. 2-9   DISPLACEMENT VECTORS.

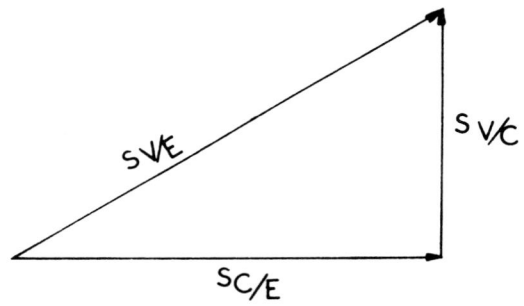

FIG. 2-10   VECTOR TRIANGLE FOR VEHICLE MOVING
ON THE FLIGHT DECK OF A CARRIER.

change in position of the point in moving from P_1 to P_2 is the vector
Δs of Fig. 2-9. The arbitrary selection of any other point for O would
not alter vector Δs.

The vector triangle of Fig. 2-9 is mathematically represented as

$$s_1 \mapsto \Delta s = s_2$$

or $$\Delta s = s_2 \to s_1$$

Suppose we place the reference point O at P_1. Then for this case $s_1 = 0$ and $\Delta s = s_2 =$ displacement.

Example. During the time period that an aircraft carrier moves 90 ft forward, a vehicle on the flight deck moves 50 ft across the deck as shown in Fig. 2-10. Determine the absolute displacement of the vehicle.

Solution. We will use the following symbols:

$s_{v/E}$ = displacement of vehicle relative to the earth (absolute displacement)

$s_{C/E}$ = displacement of the carrier relative to the earth

$s_{v/c}$ = displacement of vehicle relative to the carrier

Then

$$s_{v/E} = s_{v/c} \mapsto s_{C/E} \qquad \text{See Fig. 2-10.}$$

2.4. VELOCITY.

Velocity, a vector quantity, measures the rate of change of displacement:

$$\text{Velocity} = v = \frac{\Delta s}{\Delta t} = \frac{s_2 - s_1}{t_2 - t_1}$$

where Δs = change in displacement = $s_2 - s_1$

Δt = the change in time for Δs. $\Delta t = t_2 - t_1$

s_2 = final displacement at time t_2

s_1 = initial displacement at time t_1.

We assume a small time interval for Δt. If Δt is a relatively long time interval, then an average velocity for the time period is obtained, since the velocity could vary during a long time interval. If Δt approaches zero, then $\Delta s / \Delta t$ gives an instantaneous velocity.

Speed is a scalar quantity; it is the magnitude of velocity. The units of velocity are length per unit time such as cm/sec (cm/s) or mph.

The management of velocities is illustrated in the following examples.

Example 1. The velocity of an urban bus between stops is given by the graph of Fig. 2-11. During the first 10 seconds the velocity increases from stop to 20 fps. The bus then moves at a steady 20 fps for 20 seconds, then decelerates to a stop in 2 seconds. Determine the total distance traveled between the two stops and the average velocity of the bus.

Solution During the 10 seconds of acceleration the average velocity is 10 fps, the average of zero and 20 fps. This assumes a uniform acceleration. Then

$$s_1 = v_{\text{avg}}(\Delta t) = 10 \times 10 = 100 \text{ ft}$$

For the following 20 seconds at uniform speed

$$s_2 = v(\Delta t) = 20 \times 20 = 400 \text{ ft}$$

Finally, for the deceleration period the average velocity is 10 fps, and

$$s_3 = v(\Delta t) = 10 \times 2 = 20 \text{ ft}$$

The total displacement of the bus between stops is the sum of these three displacements:

$$\Sigma s = 100 + 400 + 20 = 520 \text{ ft}$$

The average velocity

$$= \frac{\Sigma s}{\Sigma t} = \frac{520 \text{ ft}}{32 \text{ s}} = 16.25 \text{ fps}$$

Alternately, to find the total displacement, the area under the $v - t$ curve could be calculated thus (area of two triangles and one rectangle):

$$\Sigma s = \tfrac{1}{2}(10 \times 20) + (20 \times 20) + \tfrac{1}{2}(20 \times 2) = 520 \text{ ft}$$

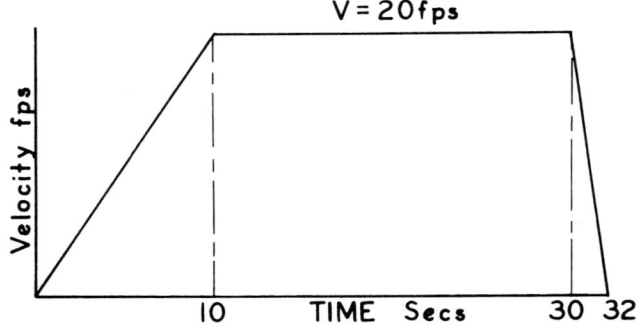

FIG. 2-11   VELOCITY GRAPH FOR A BUS.

Example 2. A small aircraft flies at a cruising speed of 240 km/h, and its intended direction is due north. Its flight path is influenced by a 55 km/h wind from the northeast. Find the actual ground speed of the aircraft and its required course.

Solution. The velocity triangle for this case of navigation is given in Fig. 2-12.

Here: v_a = actual course and velocity

v_w = velocity vector for the wind

v_s = airspeed velocity vector for the aircraft

then

$$v_s \mapsto v_w = v_a$$

The triangle must be solved by means of the sine law.

$$\frac{240}{\sin 135°} = \frac{55}{\sin \alpha}$$

$$\sin \alpha = 0.1623 \quad \text{and} \quad \alpha = 9.4°$$

Then $$\beta = 35.6°$$

$$\frac{v_a}{\sin 35.6} = \frac{240}{\sin 135°}$$

$$v_a = 197 \text{ km/h}$$

The required course is 9.4° east of north.

Example 3. At a certain instant the two aircraft of Fig. 2-13 are flying at 45° and 90° directions as shown. Displacements s_A and s_B are 5 km and 6 km, respectively, from the point of departure P. The speed of aircraft A is 300 km/h and of B 400 km/h, in the directions shown for displacement. Determine the displacement and the relative velocity of aircraft B with respect to aircraft A.

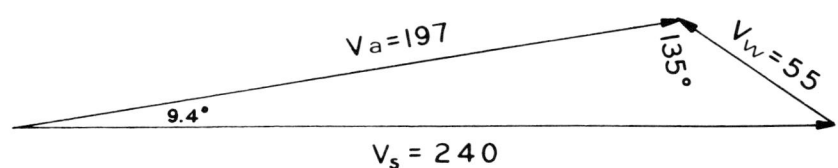

FIG. 2-12 AIRCRAFT SPEED AND COURSE ALLOWING FOR WIND.

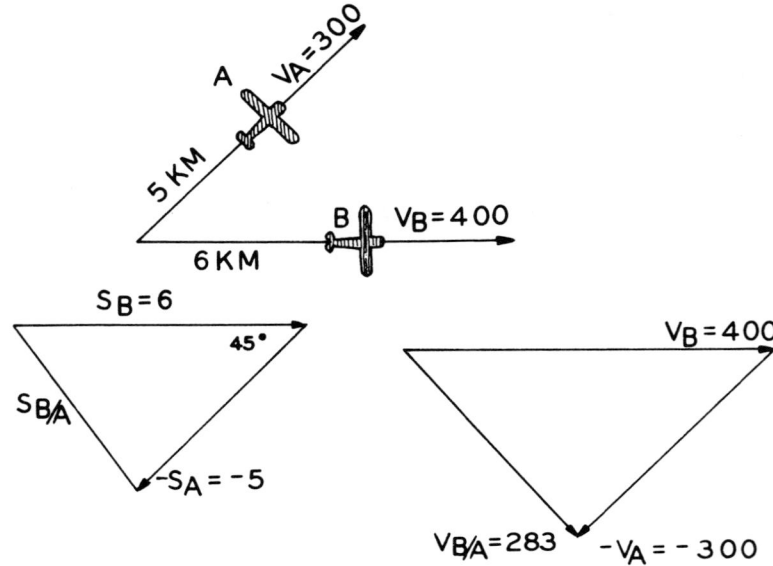

FIG. 2-13   RELATIVE MOTION OF TWO AIRCRAFT.

Solution.   For the displacement of B with respect to A

$$= s_B = s_A + s_{B/A}$$

$$s_{B/A} = s_B \rightarrow s_A \qquad \text{See Fig. 2-13.}$$

To solve this triangle use the cosine law:

$$(s_{B/A})^2 = 5^2 + 6^2 - 2(5 \times 6)\cos 45°$$

$$s_{B/A} = 4.3 \text{ km}$$

For relative velocities:

$$v_A + v_{B/A} = v_B \qquad \text{See Fig. 2-13.}$$

$$v_{B/A} = v_B - v_A$$

Solve using the cosine law as before:

$$(v_{B/A})^2 = 300^2 + 400^2 - 2(300 \times 400)\cos 45°$$

$$v_{B/A} = 283 \text{ km/h}$$

2.5. LINEAR ACCELERATION.

Acceleration is also a vector quantity. It is a change in velocity, either in magnitude or in direction. If there is no change in direction, the acceleration is linear. If the velocity changes by equal amounts in equal periods of time, the acceleration is uniform:

$$\text{acceleration} = a = \frac{\Delta v}{\Delta t}$$

$$= \frac{v_2 - v_1}{t_2 - t_1}$$

where
$$v_2 = \text{final velocity at time } t_2$$
$$v_1 = \text{initial velocity at time } t_1$$

If $t_1 = 0$, then $v_2 - v_1 = at$ if a is either a uniform acceleration or an average acceleration in the time period.

The units of acceleration are length per unit time per unit time, such as cm/s^2, ft/sec^2, etc. The scalar magnitude of acceleration is also called acceleration.

Example 1. A load is accelerated upward by a winch from zero to 600 meters/min in 5 seconds. What is the rate of acceleration?

Solution.

$$a = \frac{v_2 - v_1}{t} = \frac{\dfrac{600}{60} - 0}{5} = 2 \text{ m/s}^2$$

Example 2. A semitrailer decelerates from 60 mph to a stop in 15 seconds.

a) What is the rate of deceleration?

b) Assuming that the trailer brakes have failed and that the truck must brake the trailer, what is the force of deceleration if the total trailer weight is 40,000 pounds?

a) Deceleration:

$$a = \frac{v_2 - v_1}{t} = \frac{0 - 88 \text{ fps}}{15} = -5.86 \text{ ft/sec}^2$$

Since the acceleration is opposite in direction to the velocity, it is negative if the velocity is positive.

b)

$$\text{Force of deceleration} = F = Wa/g$$
$$= \frac{40000}{32.2} \times 5.86$$
$$= 7280 \text{ lb}$$

Forces are proportional to accelerations.

Example 3.   The connecting rod of a compressor at time t_o is moving forward at 5 cm/sec; 2 seconds later it is moving backward at 3 cm/sec. What is the average acceleration in this 2-second interval?

Solution.

$$a_{\text{avg}} = \frac{\Delta v}{\Delta t} = \frac{5-(-3)}{2} = 4 \text{ cm/s}^2$$

For cases of uniform acceleration, such as the free fall of gravitational acceleration, some useful mathematical relationships may be derived for displacement, velocity, acceleration, and time.

For uniform acceleration, the displacement is the product of the average velocity times the time interval:

$$s = \tfrac{1}{2}(v_2 + v_1)t$$

where
$$v_2 = \text{final velocity}$$
$$v_1 = \text{initial velocity}$$

If the relationship $v_2 = v_1 + at$ is substituted for v_2 in this equation, we have a useful relationship between a and s:

$$s = v_1 t + \tfrac{1}{2}at^2$$

Again, substituting

$$t = \frac{v_2 - v_1}{a} \text{ in } s = v_1 t + \tfrac{1}{2}at^2 \text{ we obtain,}$$

$$t^2 = \frac{v_2{}^2 - 2v_2 v_1 + v_2{}^2}{a^2}$$

$$s = v_1 t + \frac{at^2}{2}$$

$$= v_1 \left(\frac{v_2 - v_1}{a} \right) + \frac{a}{2} \left(\frac{v_2{}^2 - 2v_2 v_1 + v_1{}^2}{a^2} \right)$$

$$= \frac{v_2{}^2 - v_1{}^2}{2a}$$

$$v_2{}^2 = v_1{}^2 + 2as$$

Example 4. A car initially moving at 10 mps accelerates uniformly to 30 mps in 10 seconds. Determine

a)   The acceleration

b)   The distance traveled in the 10-second interval

c)   The distance traveled in the eighth second

d)   The speed after traveling 100 meters

Solution.

a)

$$a = \frac{v_2 - v_1}{t} = \frac{30 - 10}{10} = 2 \text{ m/s}^2$$

b)

$$s = v_1 t + \frac{at^2}{2} = 10 \times 10 + \frac{2 \times 10^2}{2} = 200 \text{ m}$$

c)   The distance traveled in the eighth second is found from the total distance traveled over 8 seconds less the distance traveled in the first 7 seconds:

$$s_8 - s_7 = \left[10 \times 8 + \frac{2 \times 8^2}{2} \right] - \left[10 \times 7 + \frac{2 \times 7^2}{2} \right] = 25 \text{ m}$$

d)   To find the speed at 100 meters, use

$$v_2{}^2 - v_1{}^2 = 2as$$
$$v_2{}^2 - 10^2 = 2 \times 2 \times 100$$
$$v_2 = 22.36 \text{ m}$$

Example 5. A projectile is fired vertically upward with an initial velocity of 400 mps. Neglecting air resistance, what maximum height above the firing point does the projectile reach and what interval of time is required to reach this position? Gravitational acceleration is 9.80 m/s^2.

Solution.   If we assume the upward direction to be positive, then gravitational acceleration is negative.

The maximum height occurs at that height where the instantaneous velocity is zero.

$$s = \frac{v_2{}^2 - v_1{}^2}{2a} = \frac{0 - 400^2}{-2 \times 9.80} = 8.16 \times 10^3 \text{ m}$$

To find the corresponding time of flight, use

$$s = \tfrac{1}{2}(v_2 + v_1)t$$
$$8160 = \tfrac{1}{2}(0 + 400)t$$
$$t = 40.8 \text{ sec}$$

Example 6.   An aircraft flies at 200 mps in a direction north 25° east. In a 4-second interval it turns to change course, the new course being due east at the same speed. What is its acceleration while making the turn?

Solution.   In this case the magnitude of the velocity does not change, but the direction of the velocity changes. Either a change in the magnitude or in the direction of a velocity is an acceleration.

$$v_1 \mapsto \Delta v = v_2$$

or $$\Delta v = v_2 \to v_1 \qquad \text{(See Fig. 2-14.)}$$

Using the cosine rule

$$(\Delta v)^2 = v_1{}^2 + v_2{}^2 - 2v_1 v_2 \cos 65°$$

$$= 200^2 + 200^2 - 2(200)^2 \times 0.4226$$

$$\Delta v = 214.5 \text{ m/s}$$

$$a = \frac{\Delta v}{\Delta t} = \frac{214.5}{4 \text{ sec}} = 53.6 \text{ m/s}^2$$

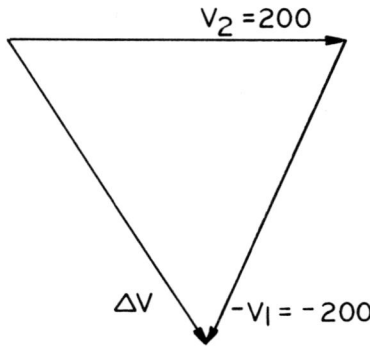

FIG. 2-14   CHANGE IN DIRECTION OF A CONSTANT VELOCITY.

2.6. PROJECTILE MOTION.

In the analysis of the trajectory (the path of motion) of a projectile two basic components of motion dominate. There are also some lesser, but still significant, effects such as air resistance, wind, and some even more complex effects. The two major components of projectile motion are:

1. The horizontal component of velocity is constant if air resistance, wind, and other effects are neglected, and is equal to the initial velocity. If the projectile is fired from a gun, the initial velocity is called the muzzle velocity.

2. The vertical component of velocity is governed by gravitational acceleration.

The combination of these two components is called *projectile motion*, and the path from the origin of motion to the point of impact is called the *trajectory*.

Fig. 2-15 shows the trajectory of a projectile with an initial velocity v_1 at an angle to the horizontal of θ. Coordinate axes are established at the point of origin. The initial components of velocity are

$$v_x = v_1 \cos\theta$$
$$v_y = v_1 \sin\theta$$

Since v_1 is constant, the X-component, or the horizontal displacement,

$$= x = v_x t = (v_1 \cos\theta)t$$

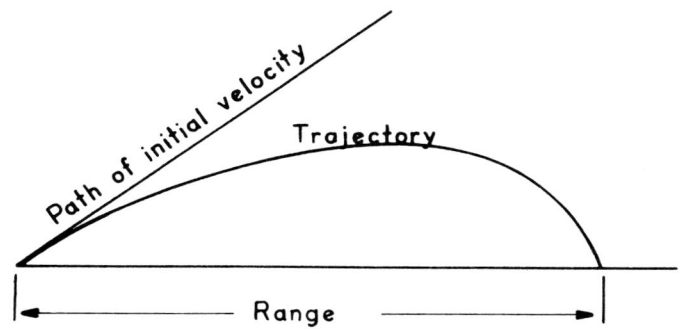

FIG. 2-15   PROJECTILE MOTION.

The Y-component of displacement includes gravitational acceleration. Recall that

$$s = v_1 t + \frac{at^2}{2}$$

so that

$$y = (v_1 \sin \theta)t - \frac{gt^2}{2}$$

To determine the velocity at any point P on the path of the trajectory

$$v_p = \sqrt{v_x^2 + v_y^2}$$

where

$v_x = X$-component of initial velocity

$v_y = v_1 \sin \theta - gt$

Example 7. A flat belt bulk material conveyor elevates material at an angle of 30° and a speed of 1.5 fps, with discharge of the material over the end pulley of the conveyor (Fig. 2-16). The point of discharge is 200 ft above grade level. Determine

a) The time of flight until material reaches the ground

b) The horizontal distance from the conveyor travelled by the first material off the conveyor

c) The velocity of the material at the end of its flight.

Solution. From Fig. 2-16 the components of velocity at the instant of discharge are

$$v_x = 1.5 \cos 30°$$
$$v_y = 1.5 \sin 30°$$

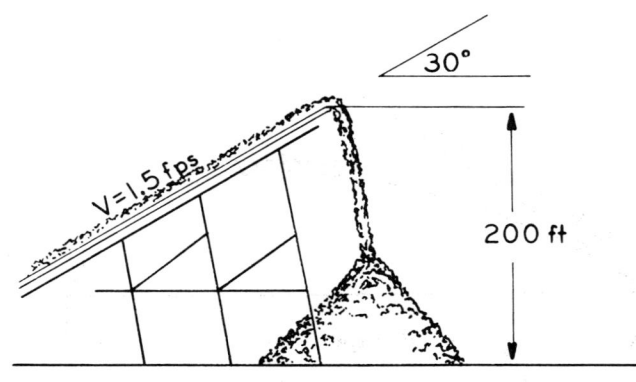

FIG. 2-16 DISCHARGE OF BULK MATERIAL FROM A CONVEYOR.

The range, or horizontal distance of flight,
$$= v_x t = 1.5(\cos 30°)t = 1.3t \text{ ft.}$$

a) The time of flight is found by using the vertical displacement of 200 ft in the equation

$$s = v_1 t + \frac{at^2}{2}$$

$$-200 = 0.7st - \frac{32.2t}{2}$$

$$16.1t^2 - 0.7st - 200 = 0$$

$$t = \frac{0.75 \pm \sqrt{0.75^2 + 4 \times 16.1 \times 200}}{32.2}$$

$$= 3.55 \text{ sec}$$

Both distance and acceleration being vertically down are taken as negative.

b) Using $t = 3.55$, the range can be found:
$$\text{Range} = 1.5 \cos 30° \, t = 1.3 \times 3.55 = 4.6 \text{ ft}$$

c) The terminal velocity $= v_2 = v_1 + at$
where $v_1 =$ initial vertical component of velocity.
Thus, $v_2 = 1.5 \cos 30° - 32.2 \times 3.55 = -113$ fps

This assumes of course that the material is falling vertically at the end of the flight, as it is. If the terminal velocity is taken (properly) as the vector sum of horizontal and vertical velocities, the result is the same:

$$v_2 = \sqrt{v_x{}^2 + v_y{}^2} = \sqrt{1.5 \cos 30° + (-113^2)}$$

$$= \sqrt{1.3^2 + 12770} = 113 \text{ fps}$$

2.7.   GRAPHICAL ANALYSIS.

Sometimes the motion of a body varies in such a manner that no convenient mathematical relationships can be found for displacement, velocity, or acceleration. Recourse must then be made to graphical methods. In Example 1 of Sec. 2.4 a graphical solution was suggested as an alternate to the mathematical solution. For a more thorough discussion of graphical analysis an example will be taken from biomechanics.

Suppose we wish to know the instantaneous velocities, accelerations, and even power output of a racing quarter horse, which happens to be the fastest breed of horse. From motion picture photography, displacement data are collected. The following figures are slightly rounded off from data produced in the *Quarter Horse Journal*, April 1978, by Dr. G.W. Pratt, Jr.

DISPLACEMENTS OF A QUARTER HORSE IN A ONE-QUARTER MILE RACE

Time, sec	Distance, ft
0.8	10
1.25	22
1.67	37
2.1	53.7
2.5	71.5
2.8	90
3.2	109
3.5	129
3.9	149
4.25	170
4.9	212
5.6	255.5
6.3	300
7.0	344
7.7	388
9.0	478
10.5	569
11.75	658
13.1	748
14.5	837
15.8	926
17.2	1014
18.6	1102
20.0	1190
21.3	1277
22.0	1320

Distance versus time is plotted on the graph of Fig. 2-17. The variation in displacement in the first few seconds is not easily determined mathematically. The horse appears to travel about 12 ft in the first second; about 41 ft (53 − 12) in the next second; about 56 ft (109 − 53) in the third second. For most of the race the rate of displacement is virtually constant.

Velocity is of greater interest and significance in races than displacement. What is the maximum speed of this horse? The maximum speed or any other instantaneous speed is easily determined from the displacement graph by recalling that

$$v = \frac{\Delta s}{\Delta t}$$

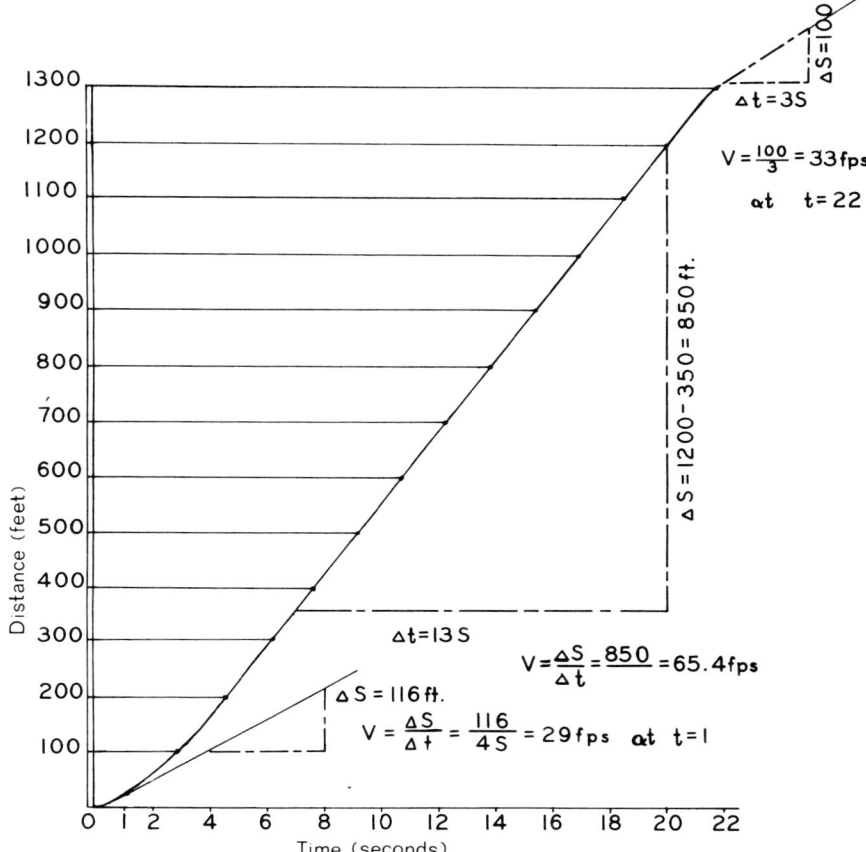

FIG. 2-17 GRAPH OF DISTANCE VS. TIME FOR A
QUARTER HORSE RACING 1/4 MILE.

That is, the slope of the displacement curve at any point of that
curve is the velocity at that instant. From 4 to 21 seconds the curve
has a uniform slope: the horse runs most of the distance at a
constant velocity. This slope is measured as

$$\frac{850 \text{ ft}}{13 \text{ sec}} = 65.4 \text{ fps or better than 45 mph}$$

We can determine by observation that this is the steepest slope and
therefore the highest speed of the horse.

The horse must begin the race from zero speed and accelerate
to its top speed. The speed of the horse 1 second out of the starting
gate is shown on the graph. This speed is obtained by drawing a
tangent to the curve at 1 second and finding the slope of this
tangent. This slope is 29 fps. The speed of the horse at the end of the

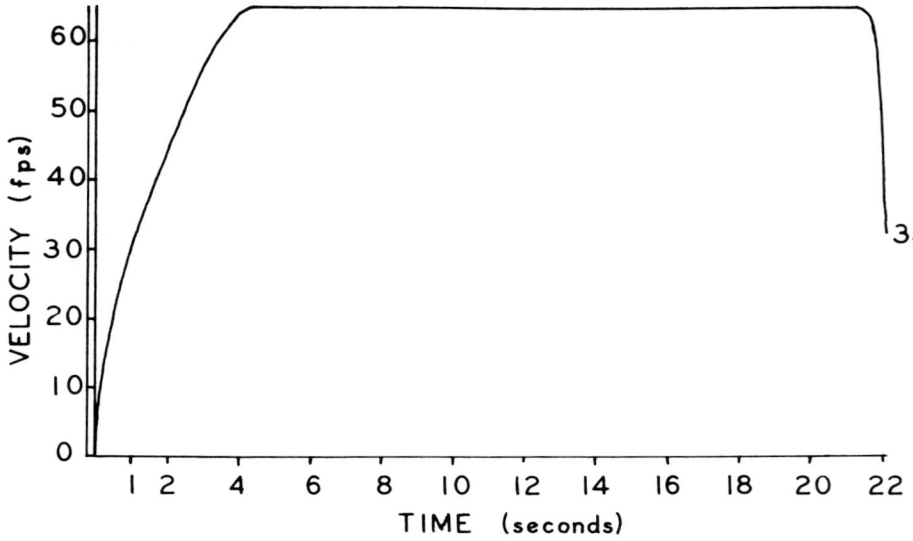

FIG. 2-18   GRAPH OF VELOCITY OF A QUARTER HORSE
RACING 1/4 MILE.

race is also shown on the graph; it is 33 fps. A plot of instantaneous
velocity versus time is given in Fig. 2-18.

Accelerations can be determined by the same method, since
$a = \Delta v/\Delta t$. A graph of velocity is plotted, then the instantaneous
acceleration at any point is found from the slope of the tangent to
the velocity curve at the point. The instantaneous slopes of the
velocity curve of Fig. 2-18 are plotted as accelerations in Fig. 2-19.
Accuracy in determining these accelerations is not possible because
of the large accelerations and decelerations that a horse can produce.
Fig. 2-19 shows a maximum acceleration rate of 40 fps^2, a formidable
figure. A calculation for the instantaneous horsepower required to
produce this acceleration in a 1000 lb horse results in a figure
exceeding 25 hp.

In this type of graphical analysis note that if the s (or v) curve
reaches a maximum or a minimum value then the v (or a) curve at
this point has a value of zero. The velocity curve of Fig. 2-18 reaches
a maximum at about 4 seconds, and at this point the acceleration
falls to zero. This would be the case, too, for a minimum value,
because the slope at a maximum or minimum point necessarily goes
through zero. If the s (or v) curve is a straight line, the v (or a) curve
will be horizontal due to the constant value of the slope of s (or v).

If the slope of a curve is steep, accuracy becomes critical.
Suppose for example that a 75° tangent is taken as 76°. Then the

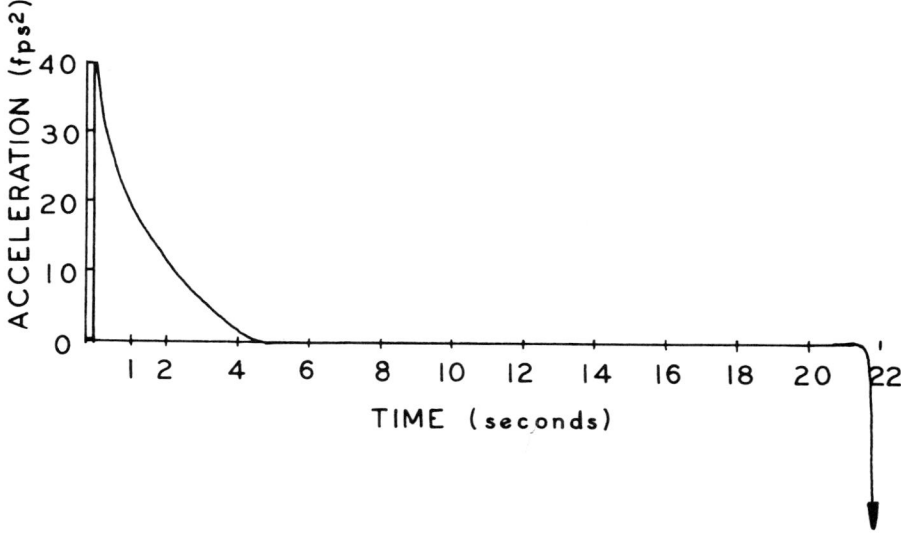

FIG. 2-19 GRAPH OF ACCELERATION OF A QUARTER HORSE RACING 1/4 MILE.

relative error

$$= \frac{\tan 76° - \tan 75°}{\tan 75°} = \frac{4.011 - 3.732}{3.732} = 7.5\%$$

The relative error is smaller for smaller angles.

The graphical analysis can be performed in reverse, that is, it is possible graphically to produce a v—t curve from an a—t curve or an s—t curve from a v—t curve.

Since $a = \Delta v / \Delta t$, then $\Delta v = a(\Delta t)$. The value of $a(\Delta t)$ is an area under the acceleration curve. This area must equal the change in velocity Δv (not the velocity, but the change in velocity). For example, consider the zero to 1 second acceleration curve of Fig. 2-19. The area under this part of the curve is approximated by a rectangle 1 second in length and a mean height of 28 fps^2. The area is 28, and checking back to Fig. 2-18 this seems to be the velocity at 1 second,—actually the change in velocity from zero velocity. Again consider the area under the acceleration curve from $t=1$ to $t=2$. This area is approximated by a rectangle 17 fps^2 high by 1 second long or 17. This is the change in velocity from 1 to 2 seconds. Since at 1 second $v=28$ fps, at 2 seconds $v=28+17=45$ fps. This figure appears to agree with the instantaneous velocity at 2 seconds given by Fig. 2-18. In the general case the change in velocity between times t_1 and t_2 is $a(\Delta t)$, and if the velocity at t_1 is v_1, then the velocity at t_2

$$= v_1 + \Delta v = v_1 + a(\Delta t)$$

Similarly $v = \Delta s / \Delta t$, so that $\Delta s = v(\Delta t)$. The value of $v(\Delta t)$ is an area under the v—t curve, and this area is the change in displacement in the time interval Δt.

The areas that are calculated may be plus or minus. If the area lies above the t axis, it is positive and if it lies below it is negative. A positive area represents a positive change in velocity or displacement; a negative area represents a negative change (decrease) in either.

PROBLEMS

1. A ship travels 50 miles in a direction south 20° west then alters course at the same speed to north 75° west. Determine the final displacement of the ship from its initial position after it has been travelling 23 miles on the new course.

2. An aircraft travels at a speed of 600 km/h in a direction 30° north of due east under conditions of no wind. Later a wind of 120 km/h from north 45° west prevails. If the aircraft does not change course, what is its new ground speed and direction under the influence of the wind?

3. Two aircraft fly together at constant speed and altitude. At a certain instant one plane maintains course and altitude at a speed of 1000 km/h while the other maintains course but drops into a power dive at 45° to the vertical at 1200 km/h. What is the magnitude and direction of the relative velocity of one aircraft to the other?

4. A body has a velocity of 40 m/s parallel to the positive X axis, then 0.8 seconds later has a velocity of 30 m/s parallel to the positive Y axis. What is the average acceleration of the body in the 0.8 second time interval?

5. A ladder is supported at 45° by a horizontal floor and a vertical wall. If the ladder slides down the wall at 1 m/s, what is the velocity of the bottom end of the ladder and the relative velocity of one end to the other end in magnitude and direction?

6. Add the four velocity vectors shown in the figure to find the resultant total velocity.

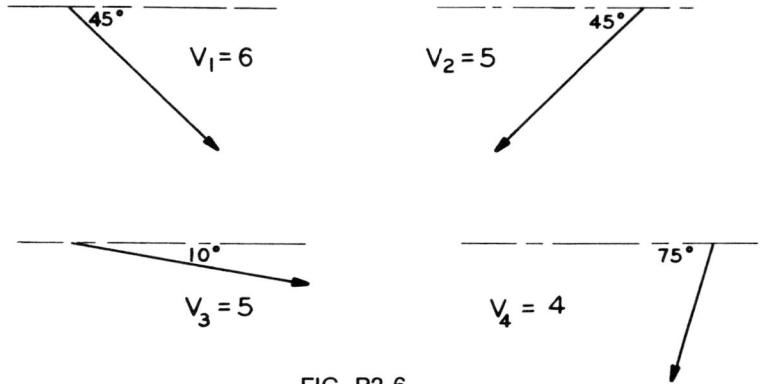

FIG. P2-6

7. A boat with a speed of 10 mph is to cross a river 2 miles wide with a current of 8 mph. Determine the velocity of the boat and the time required to make the crossing.

8. An automobile initially traveling at 10 m/s accelerates at 5 m/s² for 5 seconds. What is its speed at the end of the 5 seconds of acceleration and how far does it travel in this period of time?

9. A safety check requires that a car traveling 60 mph must be able to stop in a distance of 300 ft on dry pavement. Determine the required deceleration in ft/sec².

10. If an automobile reaches a speed of 80 km/h in 10 seconds from standstill, what is its average acceleration and the displacement in this time interval? Assume a uniform acceleration.

11. Gravitational acceleration is 9.8 m/s². Assuming no air resistance, what is the terminal velocity of a body released from an elevation of 300 meters?

12. If a baseball is thrown vertically upward with an initial speed of 40 fps, how high will the ball go and how long will it take to return to the ground?

13. During World War II, the B-17 Flying Fortress heavy bomber was used to bomb warships from high altitudes. It proved impossible for this bomber to hit a warship from high altitudes.
    a) Show mathematically why it was impossible by calculating the approximate distance a warship could move in the time required for the bomb to reach sea level. Use the following data:

bombing altitude	30,000 ft
ship speed	35 mph (about 30 knots)
aircraft speed	200 mph

    b) What was the horizontal distance traveled by the bomb in its descent?

14. Gravel is elevated 30 meters above grade by a conveyor moving at 5 m/s. Conveyor slope is 40° to the horizontal.
    a) How long does it take for the gravel to fall from the conveyor to grade level?
    b) What is the horizontal distance of travel of the trajectory of the gravel?
    c) What is the terminal velocity?

15. A bullet with a muzzle velocity of 1500 fps is fired horizontally at an elevation 5 ft above the ground. What is the time period until the bullet reaches ground level? What is its range?

16. An average speed for a baseball struck by a good hitter is 120 fps. If the ball leaves the bat at 10° above horizontal, how far does it travel to reach ground level? For convenience assume that the bat is at ground level.

17. In athletic events, a characteristic figure for the long jump by a good athlete is a takeoff speed of 30.5 fps at a 20° angle. The jump measures 26 ft 0 $-\frac{1}{2}$ in. Does the theoretical calculation for the length of trajectory agree with the measured value?

18. The acceleration of a body is 10 m/s² at time zero, decreasing uniformly to zero at $t=16$ seconds. At and after 16 seconds the acceleration is zero for 5 seconds. The initial velocity is 50 m/s and the initial displacement is zero. Draw the v—t and s—t curves for this motion.

19. A body has a constant acceleration of 5 m/s² for 3 seconds. The acceleration then decreases linearly to -3 m/s² over the next 5 seconds. At and after 8 seconds the acceleration is constant. Plot the v—t and s—t graphs for the first 10 seconds of motion. At $t=0$, $v=0$ and $s=0$.

20. A particle travels with a constant acceleration of 4 m/s² for 10 seconds. Its initial speed is -15 m/s and initial displacement is zero. Draw the v—t and s—t graphs.

21. A vehicle accelerates from stop at a rate of 4 m/s² for the first 8 seconds, moves at uniform velocity for the next 4 seconds, and then decelerates at a rate of 5 m/s². How far does the vehicle move?

22. Initial conditions for the motion of a particle are $s=0$, $v=-20$, and $a=+15$. Units are feet and seconds. At $t=5$ the acceleration of 15 falls immediately to zero. At $t=6$ linear deceleration begins, increasing to -11 at 14 seconds. Draw the a—t, v—t, and s—t graphs for this motion.

23. The s—t curve for a particle in motion can be described by the equation $s=-7+\frac{3}{4}t^2$ where $s=$ meters and $t=$ seconds. Graph the a—t curve for the period from $t=0$ to $t=5$ sec.

24. The v—t curve for the motion of a body shows a constantly increasing velocity from time zero to 60 seconds, at which time the velocity reaches 210 fps from an initial velocity of zero. At and after 60 seconds the velocity does not change

for the next 20 seconds. At 80 seconds the velocity begins to decrease at a uniform rate to zero in 30 seconds more. Draw the s—t and a—t curves for this motion.

25. A slider-crank mechanism has the following dimensions: crank length 40 mm, connecting rod 120 mm. Divide one revolution of the crank into 12 equal angular increments beginning at TDC. Calculate the displacement of the slider for each of these increments. Graph displacement, velocity, and acceleration of the slider against angle of rotation of the crank.

chapter three

Rotary Motion

3.1. ANGULAR DISPLACEMENT.

In Fig. 3-1 the weight W is suspended by a cable wrapped on a winch with a radius r to the axis of the cable. If the winch makes exactly 1 revolution to unwind the cable, then the weight is dropped by a distance $2\pi r$.

The angle through which a rotating body turns may be given in revolutions, degrees, or radians. For mathematical purposes, rotational angles are most conveniently expressed as radians. The radian is defined thus:

$$2\pi \text{ radians} = 360° = 1 \text{ revolution}$$
$$1 \text{ radian} = 57.3°$$

In one revolution any point on the winch is rotated through a distance $2\pi r$, where r is the radial distance to the point from the axis of rotation. In the general case, the distance travelled by rotation

$$= s = r\theta$$

where θ, the angle of rotation or the change in angular position, is given in radians. This equation relates linear displacement s to angular displacement θ. If $\theta = 2\pi$, then s is equal to the circumference of the circle of radius r.

61

(We have referred to a point as rotating. Since a point has no dimensions or extent, it cannot rotate. A line can rotate. In describing a point as rotating, we actually mean that a radial line on which the point is located is rotating.)

Suppose the weight in Fig. 3-1 is lowered 24 cm and the effective radius of the winch is 18 cm. What is the angular displacement?

$$s = r\theta$$
$$24 = 18\theta \qquad \theta = 0.75 \text{ radians or 43 degrees.}$$

3.2. ANGULAR VELOCITY.

Angular velocity is defined as angular displacement per unit time. Units may be revolutions per minute (rpm) or per second (rps), degrees per second, or radians per second. For computational work the radian per unit time is more convenient. Angular velocity is symbolized by ω:

$$\omega = \frac{\Delta\theta}{\Delta t}$$

If the angular velocity is variable, then to obtain an instantaneous value of ω the time interval must be small, that is, Δt should approach zero.

A point A on the surface of a rotating cylinder will have an ~~angular~~ velocity and also a linear velocity tangential to the surface

LINEAR

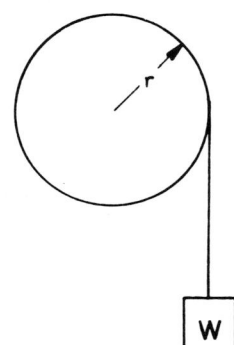

FIG. 3-1 WEIGHT SUSPENDED FROM A WINCH.

of the cylinder (Fig. 3-2). Its displacement will be

$$\Delta s = r\Delta\theta$$

and its linear velocity

$$= v_P = \frac{\Delta s}{\Delta t}$$

$$= \frac{r\Delta\theta}{\Delta t}$$

$$= r\omega$$

where ω = radians per unit time. In the general case, tangential velocity = $r\omega$.

Example 1.   A point on the circumference of a wheel of radius 0.1 meter travels through a rotation of 1 radian in 1 second at constant angular velocity. Determine ω in rpm and find the tangential velocity of the point.

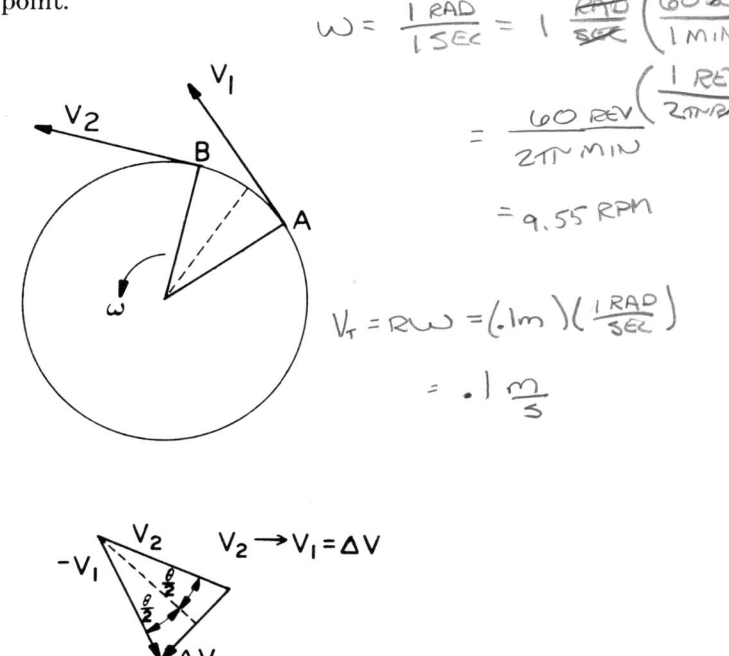

$$\omega = \frac{1\ RAD}{1\ SEC} = 1\ \frac{RAD}{SEC}\left(\frac{60\ SEC}{1\ MIN}\right)$$

$$= \frac{60\ REV\left(\frac{1\ REVS}{2\pi\ RAD}\right)}{2\pi\ MIN}$$

$$= 9.55\ RPM$$

$$V_T = R\omega = (.1m)\left(\frac{1\ RAD}{SEC}\right)$$

$$= .1\ \frac{m}{s}$$

$V_2 \rightarrow V_1 = \Delta V$

FIG. 3-2   CHANGE IN VELOCITY IN ROTARY MOTION.

Solution.

$$\omega = \frac{\Delta\theta}{\Delta t} = \frac{1\ \text{rad}}{1\ \text{sec}} = 1\,\text{r/s}$$

But 2π radians $=1$ revolution, therefore one revolution will be completed in 2π seconds.

$$\text{rpm} = \frac{60}{2\pi} = 9.6\ \text{rpm}$$

For the tangential velocity,

$$s = r\theta$$

$$= 0.1\ \text{m} \times 1\ \text{rad} = 0.1\ \text{m}$$

This displacement of 0.1 meter occurs in 1 second. The tangential velocity

$$= \frac{\Delta s}{\Delta t} = \frac{0.1\text{m}}{1s} = 0.1\ \text{m/s}$$

Example 2.   In the rolling of steel sheet, the linear velocity of the finished sheet coming out of the rolls is 1200 fpm. See Fig. 3-3. What is the angular velocity of the roll A and of the backup roll B? Assume no slippage.

Solution.   If there is no slippage, then the tangential velocities of the sheet, roll A, and roll B are all equal. But since roll B has a larger diameter than A, it will rotate more slowly.

$$v_A = v_B = 1200\ \text{fpm}$$

and
$$v_A = 2\pi r_A \omega_A$$
$$v_B = 2\pi r_B \omega_B$$

Hence
$$r_A \omega_A = r_B \omega_B$$

and
$$\frac{r_A}{r_B} = \frac{\omega_B}{\omega_A}$$

therefore
$$2\pi r_A \omega_A = 1200\ \text{fpm}$$

$$\omega_A = \frac{1200}{2\pi\frac{2}{12}} = 1146\ \text{RPM}$$

$$\omega_B = \frac{1146}{3} = 382\ \text{RPM}$$

REVOLUTION

PER

SECOND

go by Page 65

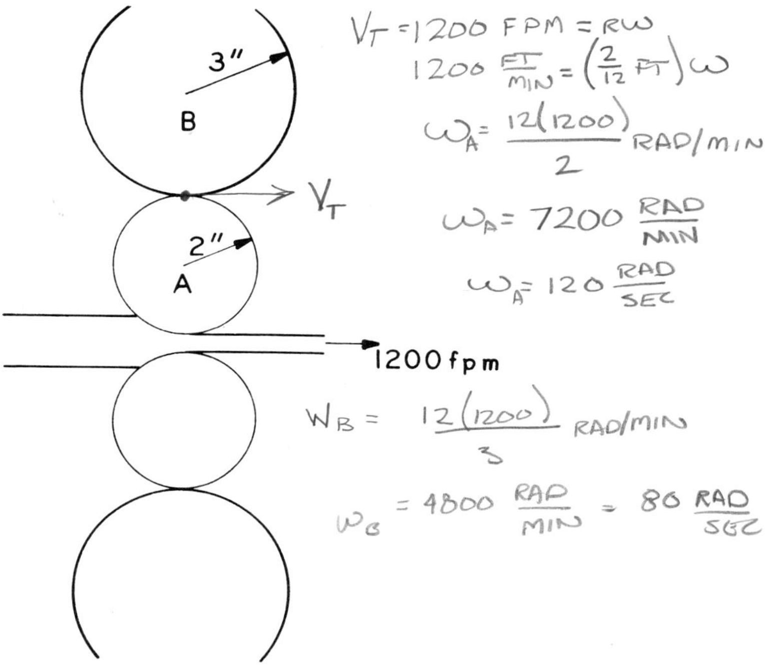

The diagram shows handwritten calculations:

$$V_T = 1200 \; FPM = R\omega$$

$$1200 \; \frac{FT}{MIN} = \left(\frac{2}{12} \; FT\right)\omega$$

$$\omega_A = \frac{12(1200)}{2} \; RAD/MIN$$

$$\omega_A = 7200 \; \frac{RAD}{MIN}$$

$$\omega_A = 120 \; \frac{RAD}{SEC}$$

1200 fpm

$$W_B = \frac{12(1200)}{3} \; RAD/MIN$$

$$\omega_B = 4800 \; \frac{RAD}{MIN} = 80 \; \frac{RAD}{SEC}$$

FIG. 3-3   ROLLING OF STEEL SHEET.

3.3.   NORMAL AND TANGENTIAL ACCELERATION.

Suppose a particle is to move in a circular path with constant angular velocity. The particle, therefore, has a constant tangential velocity. If the particle moves on the circular arc from A to B, as shown in Fig. 3-2, then there is no change in the magnitude of linear velocity, but the direction of this linear velocity changes. Either a change in magnitude or a change in direction of velocity is an acceleration, therefore the moving particle has an acceleration. It must be an acceleration without a tangential component because the tangential velocity is constant. There can be only a normal (radial) acceleration in this case.

From Fig. 3-2 $v_2 = v_1 + \Delta v$

Note that Δv is perpendicular to the chord of the circle between A and B, and the acceleration vector $\Delta v/\Delta t$ will have this direction also. The acceleration is directed toward the center of the circular path.

From Fig. 3-2
$$\frac{\Delta v}{2} = v \sin \frac{\theta}{2}$$

$$\Delta v = 2v \sin \frac{\theta}{2}$$

But for small angles in radians $\sin \theta/2 = \theta/2$

therefore
$$\Delta v = 2v \frac{\theta}{2} = v\theta$$

Also, for small angles the chord AB is closely equal to the arc AB, which is equal to $r\theta$ (θ in radians). For small time intervals the distance $AB = v \Delta t$ because AB is the distance traveled in time Δt.

$$AB = r\theta = v \Delta t$$

and
$$v = \frac{r\theta}{\Delta t}$$

Rearranging, we have $\Delta t = r\theta/v$. But $a = \Delta v/\Delta t$. Substituting $\Delta v = v\theta$ and $\Delta t = r\theta/v$ in $a = \Delta v/\Delta t$ we obtain:

$$a = \frac{v\theta}{r\theta/v} = \frac{v^2}{r}$$

This is a normal acceleration and is a characteristic of rotation. With variable rotational speed, there could also be a tangential acceleration. To differentiate between these two kinds of acceleration, subscripts are used:

$$a_n = \text{normal acceleration}$$
$$a_t = \text{tangential acceleration.}$$

An alternative form for a_n is useful if ω is known:

$$a_n = v^2/r = \frac{(r\omega)^2}{r} = r\omega^2$$

If there is a change in angular velocity ω, then there is an angular acceleration α.

$$\alpha = \frac{\Delta \omega}{\Delta t} = \frac{\omega_2 - \omega_1}{\Delta t}$$

This will give an average value of angular acceleration. If Δt approaches zero then the value for α will be an instantaneous value. Angular acceleration is a vector quantity measured in rad/sec^2.

The tangential acceleration of a particle at a distance r from the center of rotation

$$= a_t = \frac{r\omega_2 - r\omega_1}{\Delta t} = r\left(\frac{\omega_2 - \omega_1}{\Delta t}\right) = r\alpha$$

The total acceleration of a particle in rotational motion

$$= a = a_n \mapsto a_t$$

$$= \sqrt{a_n{}^2 + a_t{}^2}$$

Linear and Angular Velocity and Acceleration.

The equations for angular velocity and acceleration are of the same form as those for linear velocity and acceleration:

Linear	*Angular*
$s = \frac{1}{2}(v_2 + v_1)t$	$\theta = \frac{1}{2}(\omega_2 + \omega_1)t$
$s = v_1 t + \frac{1}{2}at^2$	$\theta = \omega_1 t + \frac{1}{2}\alpha t^2$
$v_2{}^2 = v_1{}^2 + 2as$	$\omega_2{}^2 = \omega_1{}^2 + 2\alpha\theta$

Example 1.   A point P is located on the rim of a flywheel of diameter 1 meter (Fig. 3-4). The flywheel has a clockwise angular

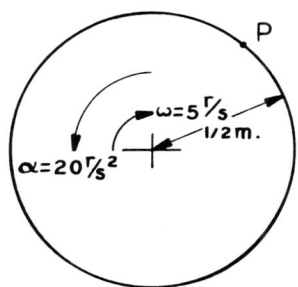

FIG. 3-4   FLYWHEEL WITH ANGULAR VELOCITY AND
ACCELERATION.

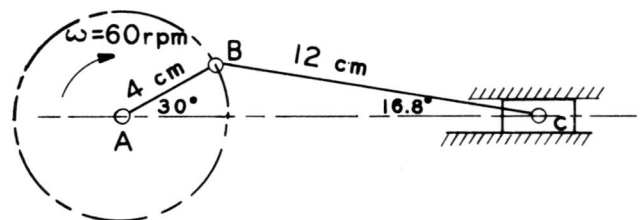

FIG. 3-5 ACCELERATION ANALYSIS FOR A SLIDER-
CRANK.

velocity of 5 rad/sec and a counterclockwise angular acceleration of
20 rad/sec^2. Determine the total acceleration of P.

Solution.

$$a_n = r\omega^2 = 0.5(5)^2 = 12.5 \text{ m/s}^2$$
$$a_t = r\alpha = 0.5(-20) = -10 \text{ m/s}^2$$
$$a = \sqrt{a_n^2 + a_t^2} = 16.0 \text{ m/s}^2$$

Example 2. In Fig. 3-5, the crank rotates with a uniform
angular velocity of 60 rpm. Determine the acceleration of the slider
C.

Solution. The motion—s, v, or a—of any point P is equal to
the motion of a point Q plus the motion of P relative to Q. In the
present case, since the motion of B is known,

$$a_C = a_B \mapsto a_{C/B}$$

In rotary motion there may be two components of acceleration,
normal and tangential, therefore, in general,

$$a_C = a_{B_n} \mapsto a_{B_t} \mapsto a_{C/B_n} \mapsto a_{C/B_t}$$

Since C moves in a straight line it does not have normal acceleration.
Also, since B has uniform angular velocity, $a_{B_t} = 0$. Then

$$a_C = a_{B_n} \mapsto a_{C/B_n} \mapsto a_{C/B_t}$$

The link BC has both linear and rotational motion, and both a
normal and a tangential component of acceleration must be assumed.

To construct the acceleration polygon, first collect the known information, which is this:

1. The normal acceleration of B can be calculated.

$$a_{B_n} = r\omega^2 = 4\left(\frac{60 \times 2\pi}{60}\right)^2 = 157.8 \text{ cm/s}^2$$

2. The direction of a_c is known. Its sense is known also. The slider is approaching TDC, where it must reverse direction. Therefore C is decelerating, that is, the acceleration of the slider is opposite, in sense, to the velocity of the slider.

3. The normal acceleration of C with respect to B is in the direction of C to B.

4. The tangential acceleration of C with respect to B is in the direction perpendicular to BC.

Thus the directions of all acceleration vectors are known. These are given in Fig. 3-6.

The normal component a_{C/B_n} can be calculated from $r\omega_{BC}^2$ or $v_{C/B}^2/r$. Since $v_{C/B}$, can be obtained from a velocity vector triangle,

$$v_C = v_B \mapsto v_{C/B}, \text{ use } a_{C/B_n} = v_{C/B}^2/r = \frac{v_{C/B}^2}{12}$$

The velocity triangle is given in Fig. 3-7. The velocity of C with

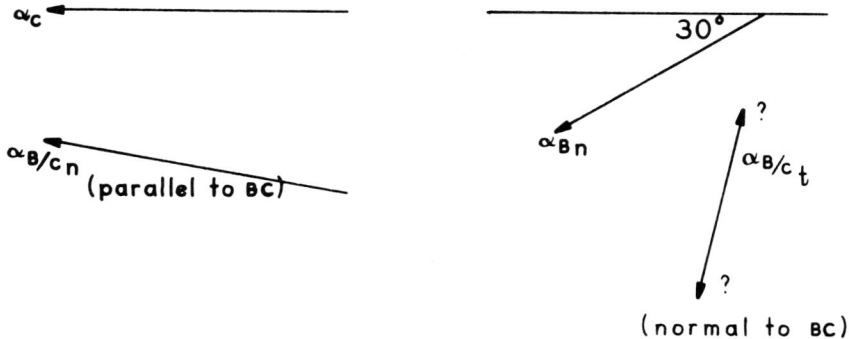

FIG. 3-6   DIRECTIONS OF ACCELERATION VECTORS IN THE SLIDER-CRANK.

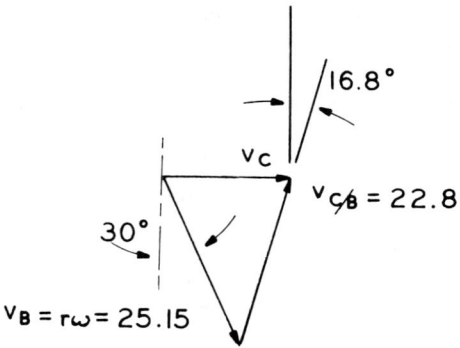

FIG. 3-7 VELOCITY TRIANGLE FOR THE SLIDER
CRANK.

respect to B must be perpendicular to BC. From the velocity triangle, $v_{C/B} = 22.8$ cm/s

Then
$$a_{C/B_n} = \frac{22.8^2}{12} = 43.2 \text{ cm/s}^2$$

This information is now put together in an acceleration polygon, Fig. 3-8. The magnitudes of a_{C/B_t} and a_C are found from their intersection in the polygon or can be found by summing X and Y components of the vectors; $a_C = 156.1$ directed to the left (deceleration).

Equations have been derived for the velocity and acceleration of the slider of a slider-crank mechanism. The following equations apply to a slider-crank for which the axis of rotation of A lies on the same plane over which C moves. If A and C are not in the same plane but offset, the mechanism is called an offset slider-crank, for which these equations do not apply.

$$v_C = (AB)\omega_{AB}\left(\sin\theta + \frac{\sin^2\theta}{2BC_{AB}}\right)$$

$$a_C = (AB)\omega_{AB}^2\left(\cos\theta + \frac{\cos 2\theta}{BC_{/AB}}\right)$$

In these equations θ is the angle of the crank from the positive X-axis. (Fig. 3-5 shows a positive θ of 30°.) The counterclockwise direction is positive. Corresponding positive directions of v_C and a_C indicate that the velocity or acceleration is toward A.

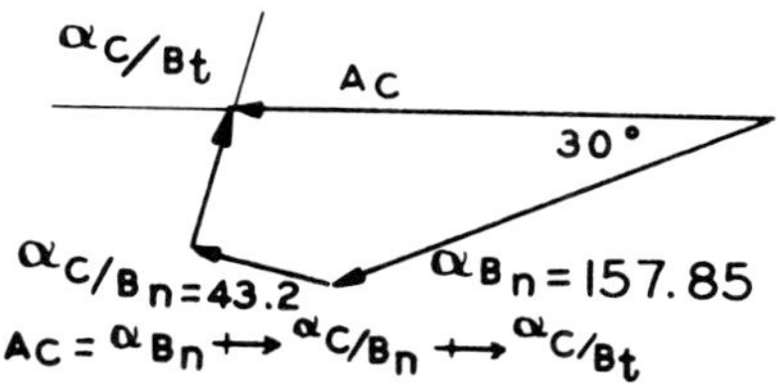

FIG. 3-8   ACCELERATION POLYGON.

For the angular velocity and acceleration of the connecting rod

$$\omega_{BC} = \frac{\omega_{AB}\cos\theta}{\left[\left(\dfrac{BC}{AB}\right)^2 - \sin^2\theta\right]^{1/2}}$$

$$\alpha_{BC} = \omega^2 \frac{\left[\left(\dfrac{BC}{AB}\right)^2 - 1\right]\sin\theta}{\left[\left(\dfrac{BC}{AB}\right)^2 - \sin^2\theta\right]^{2/3}}$$

Both ω and θ refer again to AB, not BC.

The motion of the connecting rod, and of C, can also be solved by the method of instant centers, next to be discussed.

3.4   INSTANT CENTERS.

Motion of a body in a two-dimensional plane is called plane motion. Such motion may be any of three types:

1. Translation

2. Rotation

3. Simultaneous translation with rotation, called combined motion.

A wheel rolling on a plane surface is a familiar case of combined motion. The wheel rotates and, in so doing, translates along the plane surface. The slider-crank mechanism of Fig. 1-11 illustrates all three types of motion. The slider moves with rectilinear translation; the crank rotates. The connecting rod has combined motion. Fig. 3-9 shows first its translation, then its rotation. Alternately, the link may be supposed first to rotate, then to translate. The two components of the combined motion are independent of each other.

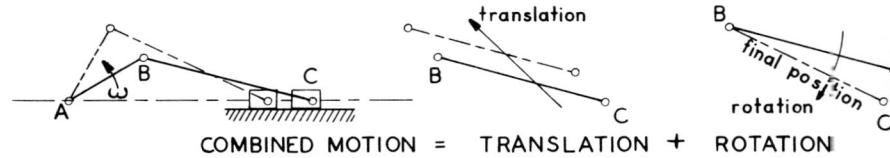

COMBINED MOTION = TRANSLATION + ROTATION

FIG. 3-9 COMBINED MOTION OF A CONNECTING ROD.

For the connecting rod, there is a unique point I about which the combined motion is reduced to pure rotation. This is shown in Fig. 3-10. This point I is called the instant center of rotation, or the centro. The connecting rod is assumed to be extended to include the instant center within its body.

An instant center is defined as the point about which a machine member rotates at a given instant. Instant centers for bodies in pure rotation, such as gears, are fixed. Others will move. In Fig. 3-11 a small block moves along a parabolic arc; its instant center is at all times the center of curvature of the arc, which is constantly changing as the curvature changes. The instant center for a body translating along a plane surface is at infinity because the radius of curvature of a flat surface is at infinity.

To locate the instant center of a body in combined motion it is necessary to know only the directions of the velocity vectors of any two points on the body. If perpendiculars to these two velocity vectors are established, then the instant center is the point of intersection of the two perpendiculars. Any point on these perpendiculars has zero velocity toward or away from the instant center. Therefore, the point at which these two perpendiculars intersect has zero velocity—the point is an instant center of pure rotation.

As an illustration, consider the case of a ladder that is sliding down a wall, Fig. 3-12. This case has no practical interest but is

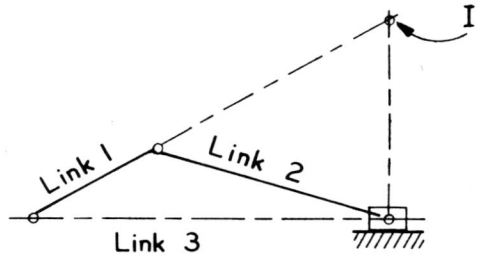

FIG. 3-10 INSTANT CENTER FOR A CONNECTING ROD.

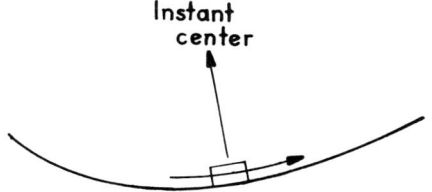

FIG. 3-11 SMALL BLOCK MOVING ON A PARABOLA.

selected because it is easy to understand. The velocity of point A, at the top of the ladder, is vertically down. The velocity of point B, at the bottom of the ladder, is horizontal. Perpendicular lines to these velocity vectors intersect at point I, therefore I is the instant center for the position shown. The instant center will change as the ladder moves. For any point P on the ladder, the velocity of P will be perpendicular to a line from P to I as shown.

If the velocity of any point on the ladder is known, such as the velocity of A, then

$$\omega = \frac{v_A}{(IA)}$$

Also $\qquad v_B = (IB)\omega \qquad$ and $\qquad v_P = (IP)\omega$

If v_A is known, then ω can be determined and the velocity of any point P can also be determined, since $v_P = (IP)\omega$. Alternately, the velocity of any point P may be found by relative velocities:

$$v_P = v_A \mapsto v_{P/A}$$

where $v_{P/A}$ must be perpendicular to the line PA.

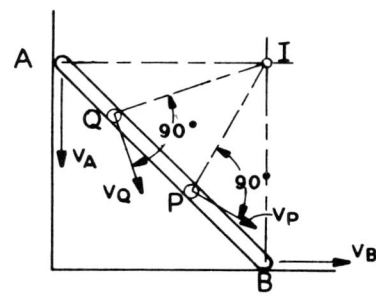

FIG. 3-12 INSTANT CENTER AND DIRECTION OF VELOCITY VECTORS.

It must be understood that this discussion of instant centers applies only to velocities and not to accelerations.

Example. Fig. 3-13 shows a slider-crank mechanism with an extended connecting rod. Determine graphically the direction of the velocity vector at D, v_D. The crank AB rotates with uniform angular velocity.

Since AB has pure rotation, v_B must be perpendicular to AB. Erect a perpendicular to v_B. This will actually be a continuation of the axis AB. The velocity of C is parallel to its fixed guides. Erect a second perpendicular to v_C. These two perpendiculars intersect at I, the instant center.

A perpendicular to the direction of v_D is given by the line ID, from which the direction of v_D is established.

In this example, the direction of the velocity of point D was determined. The concept of the instant center also makes it possible to determine the magnitude of the velocity of point D. The tangential velocity of a point on a rotating body is equal to $r\omega$, where r is

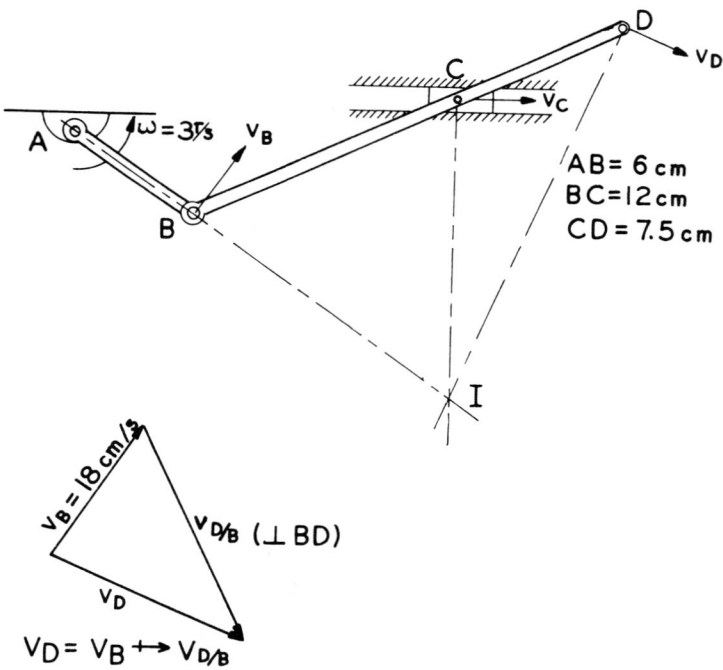

FIG. 3-13   SLIDER-CRANK WITH EXTENDED CONNECT-
ING ROD.

the radial distance from the center of rotation to the point in question. From the above example and accompanying figure, the tangential velocity of point B on AB can be calculated, since both r and ω are known for the crank AB. Since the connecting rod BCD rotates about the instant center I, the magnitude of v_D is proportional to its distance from I:

$$\frac{v_D}{\text{length } (DI)} = \frac{v_B}{\text{length } (BI)} = \omega$$

Since v_B is known and the two radial distances can be determined graphically or mathematically, the magnitude of v_D can be found.

The method of relative velocities is also available for determining v_D, since v_B is known and the directions of v_D and $v_{D/B}$ are known. The direction of v_D is found from the instant center diagram; $v_{D/B}$ must be perpendicular to the axis of BCD, as explained in Sec. 2.2. $v_D = v_B \mapsto v_{D/_B}$. The relative velocity triangle is given in Fig. 3-13.

In the four-bar linkage of Fig. 3-14 the crank AB rotates at a constant angular velocity. The angular velocity of the connecting rod BC is to be determined. This is done as follows.

Velocity v_B is perpendicular to AB. Similarly, the velocity v_C is perpendicular to the crank CD. The instant center I is located at the intersection of perpendiculars to v_B and v_C, as shown in the figure. The connecting rod BC may be considered to rotate about I at the instant that gives the position of the linkage shown.

$$v_B = \omega_{BC} \times (IB) \quad \text{or} \quad \omega_{BC} = \frac{v_B}{(IB)}$$

I is the center of rotation of the rod BC, and since all lines in a rigid

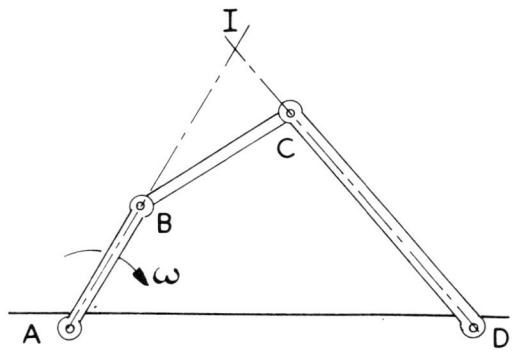

FIG. 3-14   FOUR-BAR LINKAGE WITH ITS INSTANT CENTER.

body have equal angular velocities,

$$\omega_{BC} = \frac{v_B}{(IB)}$$

Since
$$v_B = (AB)\omega_{AB}$$

$$\omega_{BC} = \frac{\omega_{AB} \times (AB)}{IB}$$

3.5.   PURE ROLLING CONTACT.

When one body rolls on another body without slipping, as occurs in the case of an automobile tire rolling on a highway surface, the two bodies are said to be in pure rolling contact. The rotating body moves by a combination of rotation and translation. Translation occurs because the center of the rotating body moves parallel to the plane surface at some linear velocity. Fig. 3-15 shows a wheel W in contact with a road surface R, with point P_w of the wheel in contact with point P_r of the road surface. If there is slippage (skidding), P_w slides and therefore has velocity while P_r is stationary. Hence, if there is slippage, the velocities of P_w and P_r differ; for pure rolling contact the velocities of these two points are identical, and if identical, then both P_w and P_r have zero velocity because P_r has zero velocity.

Now, the instant center of the wheel in combined motion is that point on the wheel, or the wheel extended in imagination, which has zero velocity. The instant center, then, is the contact point of wheel and road. Any point of the wheel has a velocity with a direction

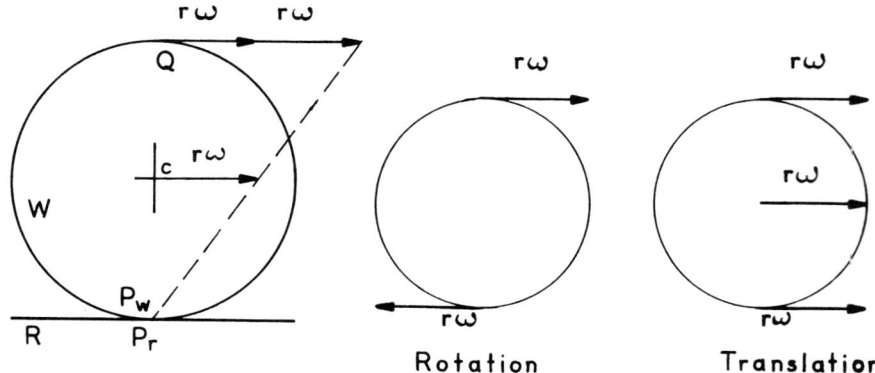

FIG. 3-15   WHEEL ROLLING WITHOUT SLIPPAGE.

which is perpendicular to a line joining that point to the instant center, the contact point. The wheel is rotating about the contact point.

In 1 minute the center of the wheel C moves $r\omega$ radians per minute or a distance $2\pi r \times$ rpm. Fig. 3-15 shows the rotational and translational velocities of the wheel. The velocities of points Q, C, and P, Fig. 3-15 are these:

v_q = velocity due to rotation $\mapsto$ velocity due to translation

$\quad = r\omega + r\omega = 2r\omega$

v_p = velocity due to rotation $\mapsto$ velocity due to translation

$\quad = -r\omega + r\omega = 0$

$v_c = r\omega$ point C translates without rotation.

The instant center of a wheel in rotation is fixed at the axis of rotation; the instant center of a rolling and translating wheel is at the point of contact with the surface on which it rolls. It is possible for the instant center of a wheel to be at some other location. One such case is given in Fig. 3-16.

The wheel in the figure is slung on a rope. One end of the rope has a velocity of 12 cm/s as shown; the other end has a velocity of 4 cm/s. The rope, therefore, is rolling on a moving surface.

The instant center of the wheel must lie on a line perpendicular to the velocities of points A and B on the wheel. The instant center then is on the line AB. But velocities must be proportional to the distances from the instant center of rotation. Since v_A is three times as great as v_B, point A must be three times as far from the instant

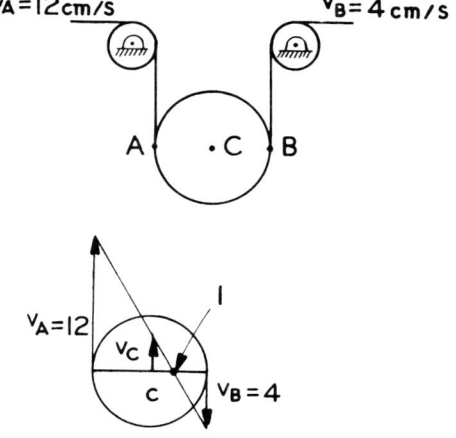

FIG. 3-16   INSTANT   CENTER   OF   A   TRANSLATING WHEEL.

center as point B. The instant center lies at a distance of one-half the radius from point B. The center C of the wheel is moving upward at 4 cm/s since C lies at the same distance from the instant center I as does B.

3.6.   KENNEDY'S THEOREM FOR INSTANT CENTERS.

In Fig. 3-14, the instant center I can be considered a momentary pivot point attached to the machine frame and about which the connecting rod rotates. That is, when the connecting rod rotates about this point, it rotates relative to the machine frame. Thus instant center I could be designated I_{23}, or I for link 2 relative to link 3.

Now consider the two disks in pure rolling contact (Fig. 3-17).

There are three members or links in this mechanism: two disks, 1 and 2, and a frame, 3. The instant center for disk 1 relative to the frame is the axis of its rotation. Designate this instant center as I_{13}, or I_{31}. (It is usual to give the smaller digit first.) Similarly the instant center of disk 2 relative to the frame is the axis of rotation. Designate this instant center as I_{23}. Disks 1 and 2 rotate relative to each other at an instant center at their point of contact, I_{12}. The three instant centers lie on one straight line. It is not a coincidence that they do.

Kennedy's Theorem for instant centers states that the instant centers of three machine members that move relative to one another must all lie on a straight line.

In Fig. 3-18 Kennedy's Theorem is applied to a four-bar linkage. Instant centers obviously common to adjacent members are labelled I_{14}, I_{12}, I_{23}, and I_{34}. I_{24} is located by the geometry of the

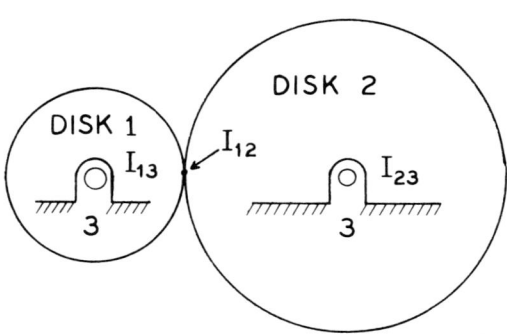

FIG. 3-17   MATING DISCS AND KENNEDY'S THEOREM.

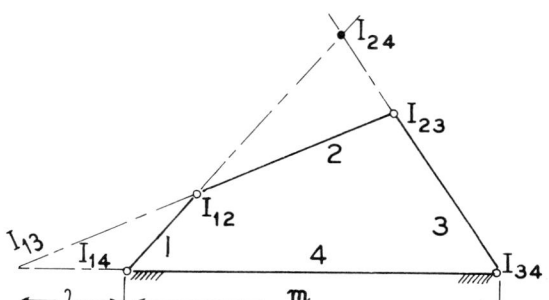

FIG. 3-18   INSTANT   CENTERS   FOR   A   FOUR-BAR
            LINKAGE.

linkage in the position shown. In accord with Kennedy's Theorem,
I_{14}, I_{12}, and I_{24} are located on the same straight line; so are I_{34}, I_{23},
and I_{24}. The geometry of the linkage locates another instant center
I_{13}; it is on the line determined by points I_{23} and I_{12}.

   Six instant centers have been located for this four-bar linkage.
Are there any more instant centers to be located? The total number
of instant centers for a mechanism is given by the following equation:

$$N_c = \frac{N_L(N_L - 1)}{2}$$

where N_c = number of instant centers
                  N_L = number of links

For the four-bar mechanism, $N_c = 4(3)/2 = 6$. All the instant centers
have been located.

   We consider next an important relationship disclosed by instant
centers. Consider for example I_{12}, the instant center located on links
1 and 2. The velocity of I_{12} on link 1 is obviously the velocity of I_{12} on
link 2. The same statement applies to, for example, I_{13}, which is a
point on links 1 and 3 if these links are extended to include I_{13}. The
velocity of I_{13} on link 1 equals the velocity of I_{13} on link 3. Since both
these links are cranks rotating about a fixed point, then

$$\omega_1 l = \omega_3 m$$

or $$\frac{\omega_1}{\omega_2} = \frac{m}{l}$$

Hence, if one angular velocity is known, the other can be determined
from the geometry of the configuration.

   For any mechanism more complex than a four-bar linkage, the
location of all the instant centers may become a confusing business.

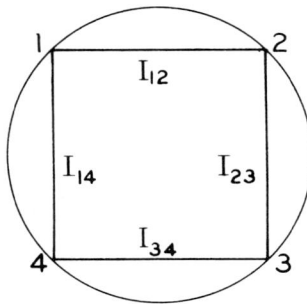 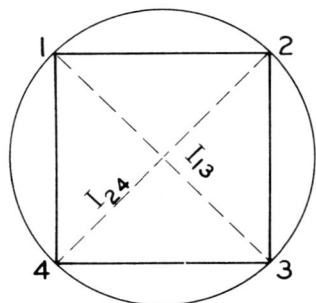

FIG. 3-19   GRAPHICAL   ANALYSIS   FOR   INSTANT
CENTERS.

If so, it is customary to draw upon a simple graphical device to provide a systematic procedure for locating and counting instant centers. The method is easily understood from the four-bar linkage of Fig. 3-19.

Draw a circle. On the circumference mark as many equally spaced points as there are links in the mechanism. In the case at hand this is 4. Number the points 1, 2, 3, etc. to designate the links. Draw solid lines between numbers for the obvious instant centers. From Fig. 3-18 instant centers 14, 12, etc. are immediately known, and solid lines are drawn for these in Fig. 3-19. Where no line exists between any two numbers, draw dashed lines to represent instant centers still to be located.

3.7.   ROLLING CYLINDERS AND CONES.

The simplest type of rolling contact is that of a cylinder or disk rolling on a plane, as discussed in the previous section. The plane may be considered as a cylinder with its center at infinity.

The two cases of cylinders in pure rolling contact are shown in Fig. 3-20. In the first case the cylinders rotate in opposite directions. If the two cylinders must rotate in the same direction, then one must be an internal cylinder as in the figure.

An essential principle of rolling contact can be observed from these cases: rolling contact occurs only when the point of contact of the two rolling bodies lies on the line of centers of the two bodies. The line of centers is the line passing through the rotational axes of the two bodies.

With pure rolling contact, that is, without slipping, the lengths of arc on the two cylinders making contact in a given interval of

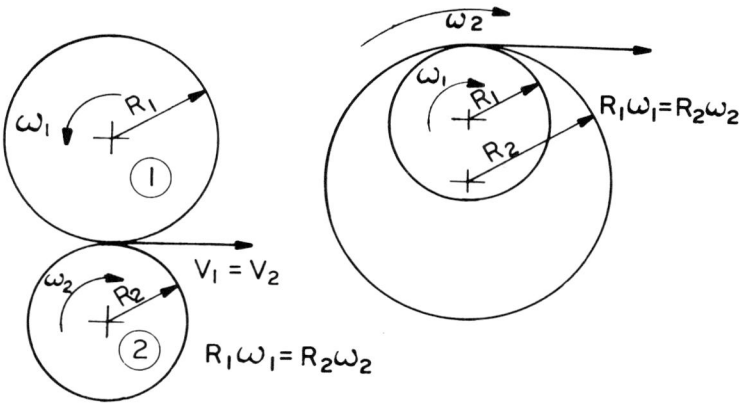

FIG. 3-20   PURE ROLLING CONTACT OF CYLINDERS.

time are equal. The center distance equals the sum of the two radii. The ratio of angular velocities

$$= \frac{\omega_2}{\omega_1} = \frac{R_1}{R_2}$$

Rolling cones and bevel gears are used to transmit rotational motion between shafts with intersecting axes as in Fig. 3-21. For pure rolling contact at T, T must lie on the line of centers, as for cylinders, and the angular velocity ratio

$$= \frac{\omega_2}{\omega_1} = \frac{TA}{TB}$$

Rolling contact occurs at all other points along the line of contact,

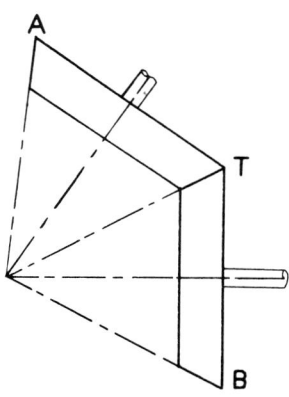

FIG. 3-21   CONES IN ROLLING CONTACT.

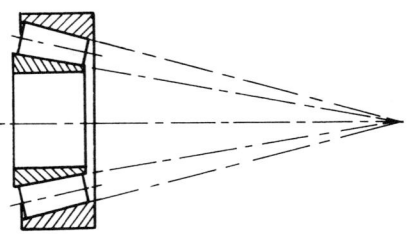

FIG. 3-22   BEARING WITH TAPERED ROLLERS.

provided that the two cones have a common apex. The case of two cylinders in rolling contact may be considered as a special case of rolling cones, with a common apex at infinity.

3.8.   ROLLING CONDITIONS IN BALL AND ROLLER BEARINGS.

Contrary to what is usually thought, pure rolling contact, without sliding, is not usual in ball and roller bearings. Because there is sliding, these bearings require lubrication. The tapered roller bearing of Fig. 3-22, however, has pure rolling contact.

Ball bearings have one or two rows of hardened steel balls positioned between raceways. The balls are usually supported in a separator between the outer and the inner ring. Roller bearings use the same general construction. The roller bearing makes contact along a line of contact, while a ball bearing has a contact area not greatly larger than a point; there is a minute flattening due to a very large compressive stress.

To understand why there is sliding in a ball bearing, refer to Fig. 3-23. The surface speed of a point near the outside diameter of the inner race is $\pi D_1 \times$ rpm. The speed at the bottom of the groove in this race is less because the diameter D_2 is less. In the case of the ball, the maximum surface speed is the ball rpm $\times \pi d_2$, but at a point d_1 the diameter is smaller and the surface speed lower. Thus, a slower part of the ball is in contact with the fastest moving part of the race, and therefore there must be sliding. To reduce sliding contact the manufacturer makes the groove in the race with a radius about 4% larger than that of the ball.

3.9.   SLIDING CONTACT.

Suppose the two bodies shown in Fig. 3-24 do not roll on one another but instead have sliding motion.

The contact point P is a point both on body 1 (P_1) and on body 2 (P_2). The velocity of P_1 relative to P_2 must be in the direction of the

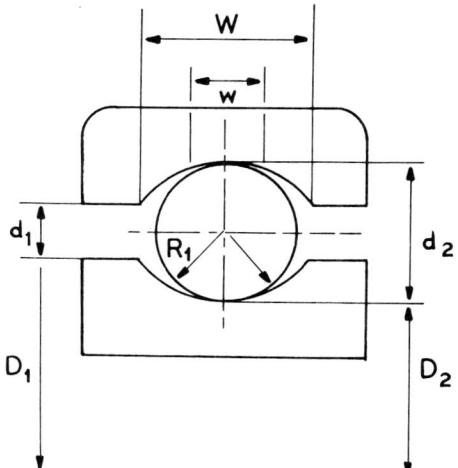

FIG. 3-23   ROLLING MOTION IN A BALL BEARING.

common tangent at point P because otherwise the two bodies would separate (or overlap). The velocity of P_1 relative to P_2, or $v_{1/2}$ is equal in magnitude but opposite in direction to the velocity of P_2 relative to P_1, or $v_{2/1}$. Both are colinear with the common tangent line. *A sliding velocity is a tangential relative velocity of the two contact points.*

If two bodies move relative to one another, they must either be. in pure rolling contact or else in sliding contact. If both bodies have identical velocities at the point of contact between them, then they are in pure rolling contact. But if the velocities of the contact points on each body differ—in magnitude, direction, or sense—then they are in sliding contact. Sliding contact is frequently found in cam mechanisms, as in the following example.

The flat-face follower F of Fig. 3-25 is driven by cam C. The face of the flat follower is colinear with the tangent line to the two bodies, and the motion of the follower is normal to this tangent line.

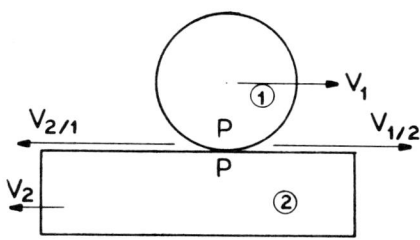

FIG. 3-24   SLIDING CONTACT.

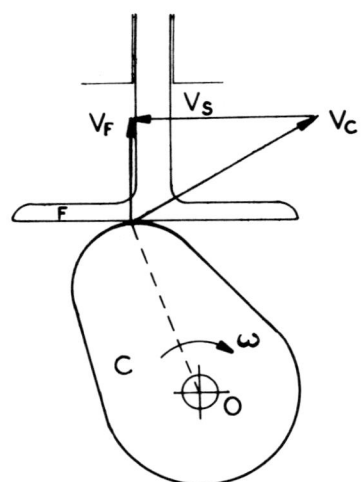

FIG. 3-25 SLIDING CONTACT BETWEEN CAM AND FOLLOWER.

The cam rotates with uniform angular velocity. The velocity of the contact point P on the cam surface is perpendicular to the line joining P to the axis of rotation. It is the velocity vector v_C in the figure. Since the cam and its follower are constrained by gravity, springs, or other devices, to maintain contact the normal component of v_C must be equal to the normal velocity of the follower, v_F. The difference between v_C and v_F is v_S, the velocity of sliding.

Is the determination of the sliding velocity of a cam an academic exercise or does it have any practical significance?

One of the major preoccupations in engineering design is wear in machine parts. A part that does not slide should not wear. Wear will be approximately proportional to average sliding velocity times time, or total displacement. The life of the sliding part will be longer if the sliding velocity can be reduced. If the cam surfaces in the example above must maintain their original accuracy of manufacture, or if the cam surfaces must have a long life, then the sliding velocity—its average and its maximum and minimum values—must be examined. Clearly, for the cam of Fig. 3-25 the sliding velocity will vary over one rotation of the cam, and, therefore, wear will be uneven on both cam and follower surfaces.

While most rolling bodies are circular, some are noncircular such as rolling ellipses and hyperboloids which will be discussed later. The oval cam with a roller follower of Fig. 3-26 would seem to be a case of rolling contact, but there is sliding also.

The cam, like most cams, rotates with uniform angular velocity.

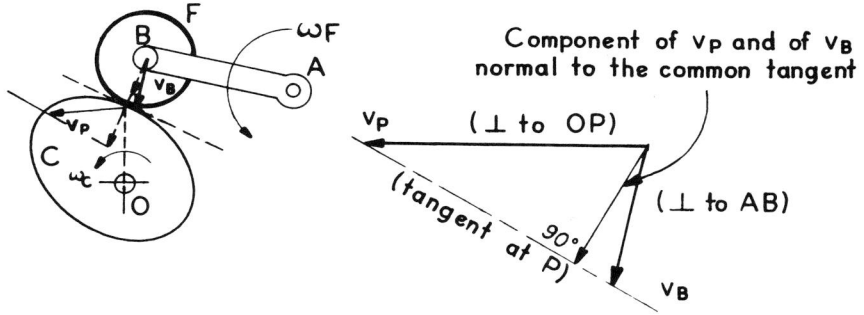

FIG. 3-26   CAM AND FOLLOWER. THE CONTACT POINT
   IS DESIGNATED P IN THE TEXT.

The contact point between roller and cam will be designated P. The instantaneous velocity of point P on the cam is found as usual by $\omega_C(OP)$. This velocity is perpendicular to the line OP. For pure rolling contact, the velocity of P on the follower must equal the velocity of P on the cam. This cannot be, if only because the velocity of P on the follower cannot have the direction of P on the cam shown in the figure. The component of the velocity of P (and of B) on the follower normal to the tangent line is equal to the component of v_P on the cam in the same direction. This must be so if cam and follower are to maintain contact. Because the follower is pivoted at A, it is moving downward as the cam rotates counterclockwise. The sliding velocity is the vector difference between the vectors v_P and v_B shown in the figure.

If the angular velocity of the arm AB must be known, it can be found from v_B/AB.

3.10.   SPECIAL CURVES USED IN MECHANISMS.

Some special curves related to the circle and their applications are discussed in the following sections. The cycloid finds most of its applications in cam design; the epicycloid is used for generating straight-line motion. The epitrochoid, a special case of the epicycloid, is used in engines and blowers such as the Wankel engine. The involute is the standard profile for gear teeth. More extended practical discussion of these curves will be found in the following chapters.

3.11.   THE CYCLOID.

The curve of the cycloid is described by successive positions of a point on the circumference of a circle that rolls on a straight line. The layout of such a curve is shown in Fig. 3-27. The generating

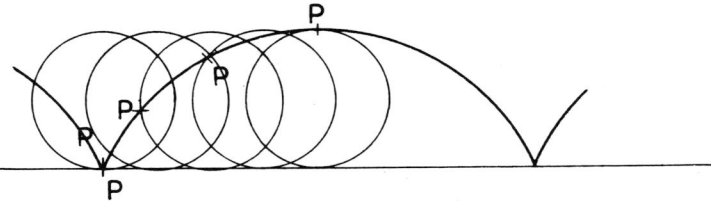

FIG. 3-27   POINT *P* DESCRIBES A CYCLOID.

circle rolls horizontally on the straight line. Successive positions of the selected point on the generating circle are labelled *P*.

The points on a cycloidal curve can be determined by the *X* and *Y* coordinates:

$$x = R(\theta - \sin\theta)$$
$$y = R(1 - \cos\theta)$$

where

R = radius of the generating circle
θ = angle in radians through which the circle is rolled.

The origin of coordinates is the starting point of the curve at contact with the straight line.

A cycloidal curve of a different path is traced by a point on the generating circle if the generating circle is rolled on the periphery of a fixed circle instead of a flat plane. The center of the generating circle will, of course, trace a circle with a radius equal to the sum of the radii of the two circles. If the generating circle has a radius one-half that of the fixed circle, a point on its circumference will trace a kidney-shaped curve.

The generating circle may be rolled on the inside of a fixed circle as in Fig. 3-28. Here *G* is the generating circle and *F* the fixed

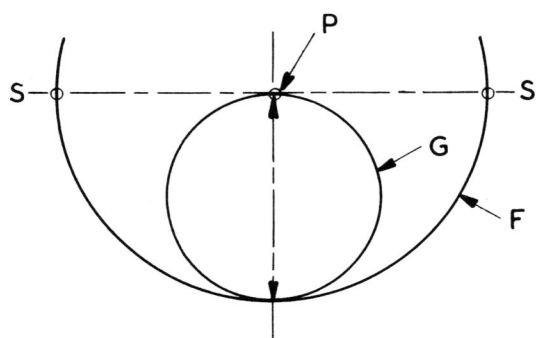

FIG. 3-28   HYPOCYCLOID.

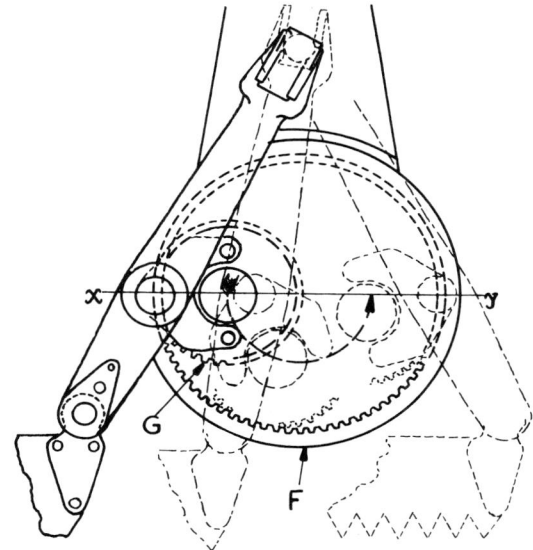

FIG. 3-29   THE HYPOCYCLOID APPLIED TO A RECIPRO-
CATING SAW.

circle. The mechanically useful applications require that the radius of
the fixed circle be twice that of the generating circle. Given this
condition, a point P on the generating circle describes the straight
line SS as the generating circle rolls; this straight line is called a
hypocycloid. The hypocycloid principle has found many applications
in straight-line motion. The saw-reciprocating mechanism shown in
Fig. 3-29 uses the principle of the hypocycloid, with, of course, gears
instead of rolling disks. In the figure G is the generating circle or
gear and F the fixed gear. The point moving in straight-line motion
is designated P, moving along line $x - y$.

3.12. THE EPITROCHOID.

The curves discussed have been generated by a point located on
the periphery of the generating circle. If, however, the tracing point
is not at the periphery, but between the center and the periphery of
the generating circle, a different type of curve, called a *trochoid*, is
traced.

Suppose the generating circle is to roll on the outside of a fixed
circle and consider the path traced by various points on the generating circle. The center of the generating circle will, of course, travel

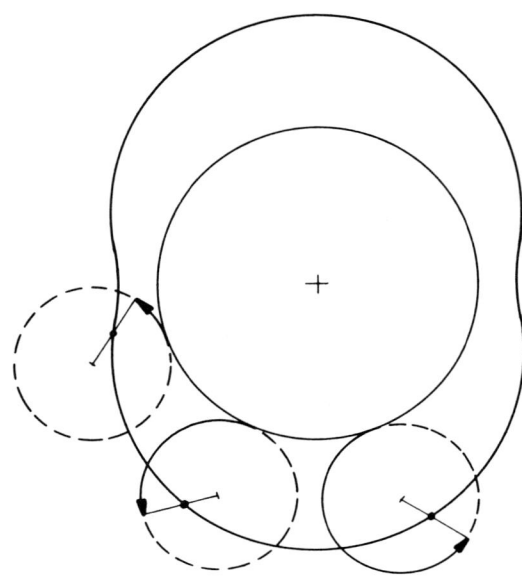

FIG. 3-30   TWO-LOBE EPITROCHOID. THE GENERATING
CIRCLE HAS HALF THE RADIUS OF THE
FIXED CIRCLE.

along a path which is another circle. If any other point on the
generating circle is used, then its path is not circular, but has a shape
called an *epitrochoid*. If the generating circle rolls around the inside
of a base circle, then any point on the generating circle travels a
curved path called a hypotrochoid. In Fig. 3-30 a generating circle
with a radius one-half that of the fixed circle is used. The epi-

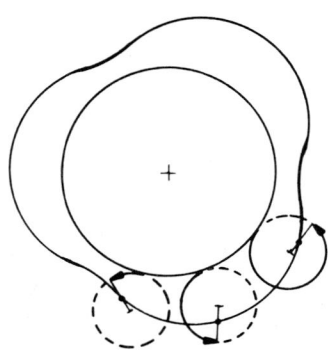

FIG. 3-31   THREE-LOBE EPITROCHOID PRODUCED BY A
GENERATING CIRCLE WITH ONE THIRD THE
RADIUS OF THE FIXED CIRCLE.

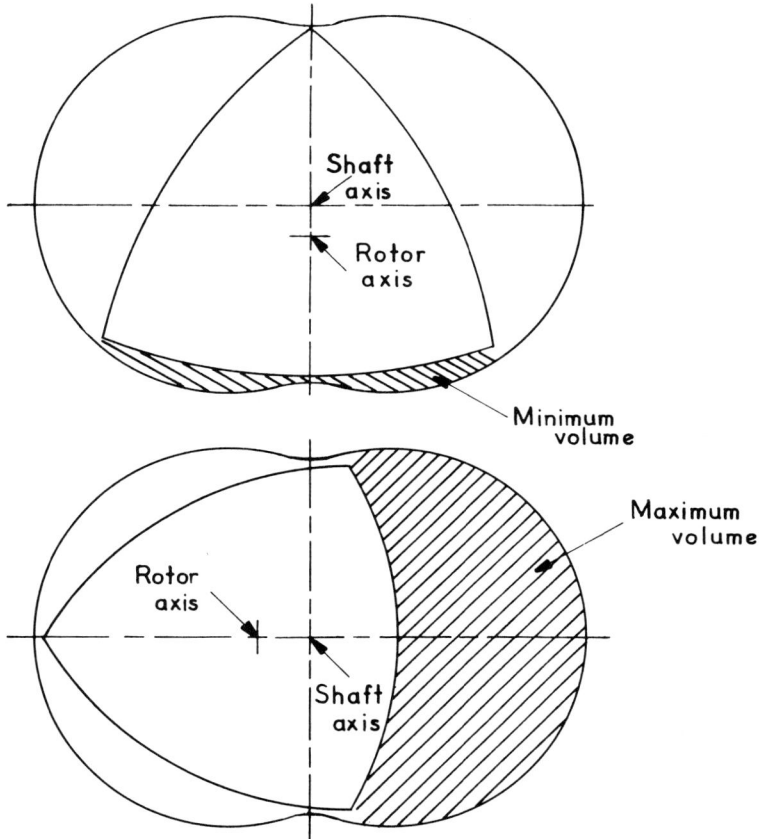

FIG. 3-32   WANKEL ENGINE ROTOR IN POSITIONS OF
MINIMUM AND MAXIMUM CHAMBER
VOLUME.

trochoidal path travelled by a point lying midway along the radius is
shown in the figure. This path is the shape of the chamber of the
Wankel internal-combustion engine. The minor axis is 2.5 times the
radius of the fixed circle and the major axis is 3.5 times the radius of
the fixed circle.

   If the generating circle has a radius one-third that of the fixed
circle, the three-lobe epitrochoid of Fig. 3-31 results. However the
minimum number of lobes, two, is always selected for the design of
mechanisms.

   The rotor of a Wankel engine requires one lobe more than the
number of lobes of the housing, and is mounted eccentrically on its

shaft. The maximum engine volume enclosed by any rotor face in this engine occurs when the apex of any rotor lobe lies on a minor or major axis of the epitrochoid (Fig. 3-32).

3.13. THE INVOLUTE.

The curve of an involute is exemplified by the free end of a tight cord as it is unwrapped from a stationary cylinder. See Fig. 3-33. The radius of curvature of the involute at any point is equal to the length of the cord that has been unwrapped. The involute is the curve used for the profile of almost all gear teeth.

The coordinates of any point of the involute may be found from the equations

$$x = R(\cos\theta + \theta\sin\theta)$$
$$y = R(\sin\theta - \theta\cos\theta)$$

where
R = radius of the cyclinder
θ = angle in radians through which the cord is unwrapped.

The origin of coordinates is the center of the circle, and these coordinates apply if the curve begins at the initial point shown in the figure.

3.14. SIMPLE HARMONIC MOTION.

Simple harmonic motion is one of the commonest and most useful types of motion or vibration. Since it is related to rotation at a uniform angular velocity, it has the virtues of mechanical simplicity

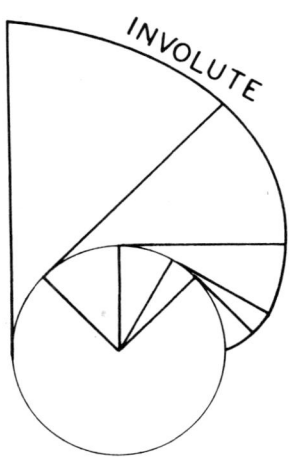

FIG. 3-33   THE INVOLUTE.

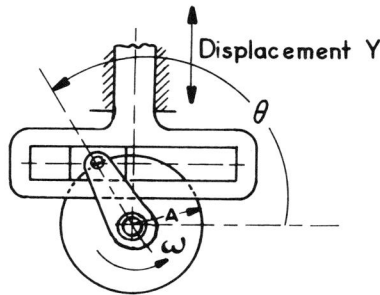

FIG. 3-34   SCOTCH YOKE MECHANISM.

and reliability. It is explained here in terms of one of its applications, the Scotch yoke mechanism of Fig. 3-34, which converts rotary motion into periodic linear motion.

The rotating hub carries a pin which engages the slot in the frame; thus the frame reciprocates up and down. There is a cycle of reciprocation for each revolution of the hub. Therefore, the period of the motion is the time required for one revolution:

$$\text{Period} = T = \frac{\theta}{\omega} = \frac{2\pi \text{ rad}}{\omega \text{ rad/sec}} = \frac{2\pi}{\omega} \text{ sec/cycle}$$

where ω = angular velocity of the hub in radians per second.

The frequency of the motion is the reciprocal of the period:

$$f = 1/T = \frac{\omega}{2\pi} \text{ cycles/sec.}$$

The vertical motion of the yoke is controlled by the pin. Thus the yoke and the pin have the same vertical displacement y, the same vertical velocity v_y, and the same vertical acceleration a_y. Expressions for displacement, velocity, and acceleration are found as follows.

The angular displacement of the pin

$$= \theta = \omega t \qquad \text{where } t = \text{time}$$

Now, if A is the amplitude of the vibration (it is also the radial distance of the pin from the axis of rotation), then

$$y = A \sin\theta \qquad \text{(See Fig. 3-35.)}$$
$$= A \sin(\omega t)$$

For vertical velocity

$$v_y = A\omega\cos\theta$$
$$= A\omega\cos(\omega t)$$

Finally

$$a_y = -A\omega^2\sin\theta$$
$$= -A\omega^2\sin(\omega t)$$

Also $a_y = -\omega^2 y = -ky$ where k is a constant, since ω is constant. This

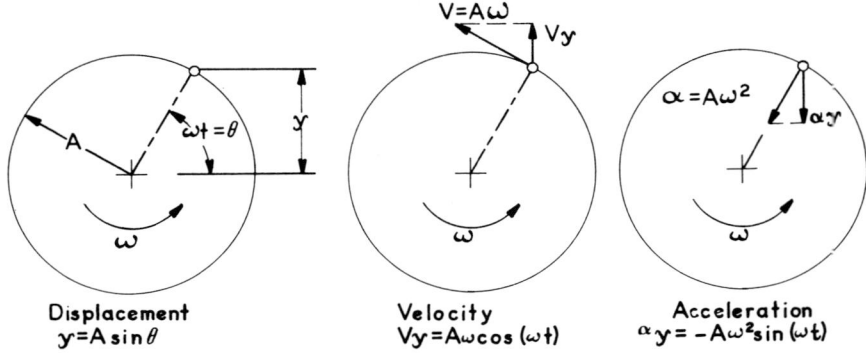

Displacement
$y = A \sin \theta$

Velocity
$V_y = A\omega \cos(\omega t)$

Acceleration
$\alpha_y = -A\omega^2 \sin(\omega t)$

FIG. 3-35   DISPLACEMENT, VELOCITY, AND ACCELERA-
TION IN SIMPLE HARMONIC MOTION.

is the defining equation of harmonic motion, that is, the magnitude
of the acceleration is proportional to the displacement and is opposite
to the displacement.

The acceleration can be found from the frequency thus:

$$f = \frac{\omega}{2\pi}$$

and
$$\omega^2 = 4\pi^2 f^2$$

Substituting this value of ω^2 into $a_y = -\omega^2 y$ gives

$$a_y = 4\pi^2 f^2 y$$

Displacement, velocity, and acceleration have similar sine
curves, but with a phase displacement to each other.

3.15.   THE SWASH-PLATE MECHANISM.

The swash-plate reciprocating mechanism is illustrated in Fig.
3-36. Perhaps its most important application is in axial piston hydraulic pumps, where it provides the reciprocating motion for the
pump pistons. The swash plate is angled to its shaft, which turns at a
constant angular velocity. The driven follower translates axially and
must be loaded by a spring or other means so that it is held against
the swash plate. The stroke of the follower can be adjusted by
changing the angle of the swash plate; a maximum angle of about
20° is used. In a swash-plate piston pump the oil delivered by the
pump is proportional to the swash-plate angle.

To analyze the motion of the follower, consult Fig. 3-37. The
motion is not uniform. The circle of the swash plate is shown in the

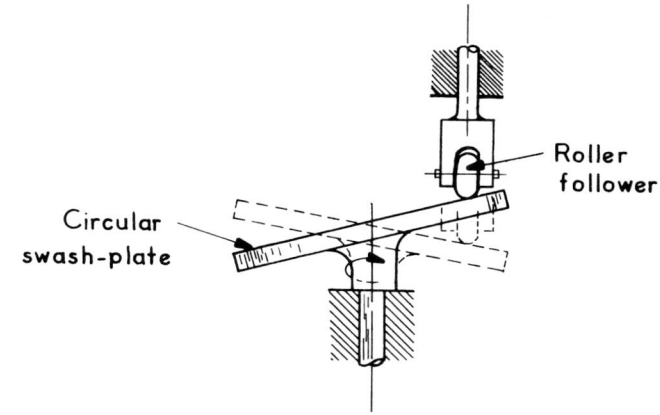

FIG. 3-36   SWASH-PLATE MECHANISM.

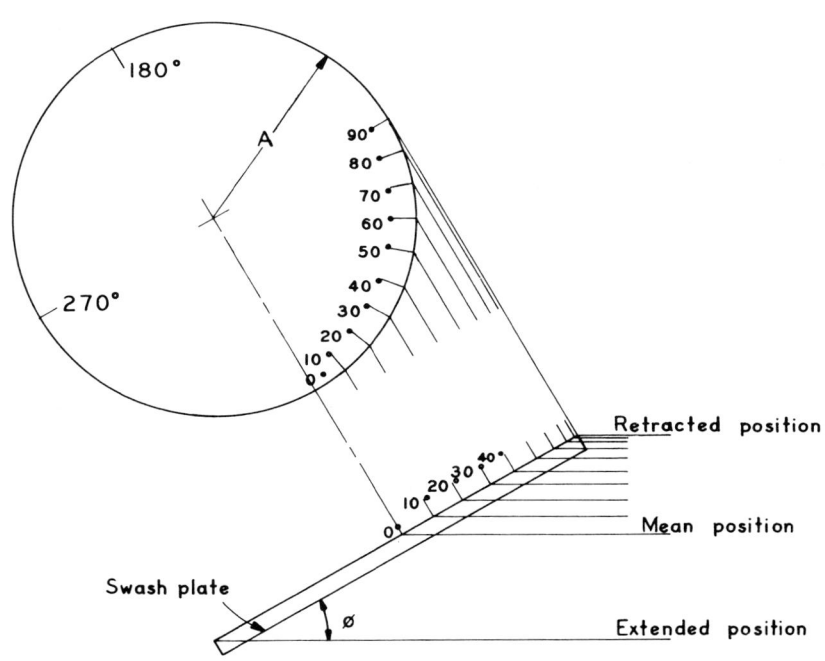

FIG. 3-37   ANALYSIS   OF   THE   SWASH   PLATE
         MECHANISM.

figure, with angles in one quadrant marked. The displacement at 10°
increments is marked on the edge of the swash plate. At zero and 180
degrees the follower displacement is zero, and the displacement is
maximum at 90 and 270 degrees.

Examination of Fig. 3-37 will disclose that the follower motion
is a case of simple harmonic motion. The motion would be the same if
the swash plate were fixed in position and the follower moved in a
circle around the plate.

3.16.   CORIOLIS ACCELERATION

Coriolis acceleration is an acceleration component that occurs
when a point follows some path, straight or curved, on a rotating
body. Consider the case in Fig. 3-38. The disk D rotates with a
constant angular velocity ω about its axis O. A block slides in the
straight radial slot. The block is sliding outward in the slot at a
constant linear velocity.

Consider two coincident points A and B at a radial distance r
from O. Point A is on the disk directly underneath point B on the
sliding block. It is required to find all the acceleration components of
B, and in the usual case $a_B = a_A \mapsto a_{B/A}$.

The velocity of $A = v_A = r\omega$, and this velocity is normal to OA.
The acceleration of $A = a_A = r\omega^2$, and this acceleration is directed
toward O, the center of rotation.

Since B is moving radially out at a constant velocity relative to
A of $v_{B/A}$, then $v_B = v_A \mapsto v_{B/A}$.

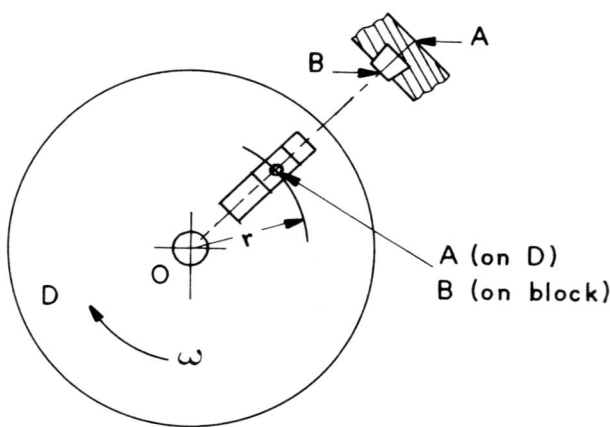

FIG. 3-38   ROTATING DISK WITH RADIALLY MOVING
BLOCK.

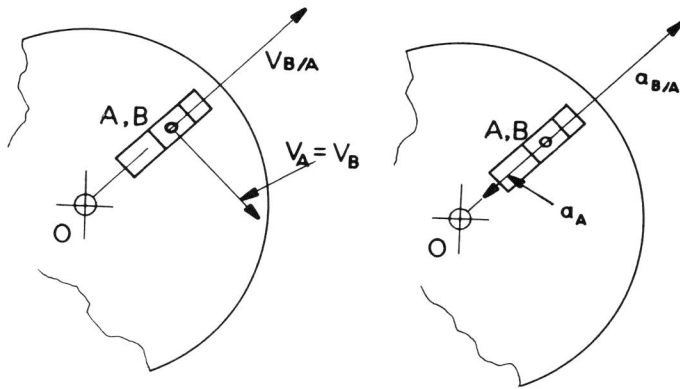

FIG. 3-39   VELOCITIES AND ACCELERATIONS FOR
BLOCK B.

The velocity and acceleration components given above are shown in Fig. 3-39. But these are not the only components of velocity and acceleration. We have neglected two considerations that apply to the movement of B:

1.   the change in direction of $v_{B/A}$ caused by the rotation of D;

2.   the change in magnitude of the tangential velocity of $B, v_B{}^T$, caused by its increasing radial distance from O. That is, $v_B{}^T = r\omega$, with r constantly increasing with time.

Since in both of these omissions there is a velocity change, we have neglected two accelerations for B.

For the change in direction of $v_{B/A}$, see Fig. 3-40. The direction of $v_{B/A}$ goes from OA to OA'. Since we are considering the rotation of D and not the radial movement of B, points A and B are coincident and so also are A' and B'.

In a small time interval Δt this change in direction is $d\theta$, and the distance traveled from A to A' is equal to $r\Delta\theta$. For small angles $r\Delta\theta$ is closely equal to the chord AB, and this is the angle turned by vector $v_{B/A}$. The change in $v_{B/A}$ due to rotation is shown in Fig. 3-40 as Δv. Comparing similar triangles

$$\frac{r\theta}{r} = \frac{\Delta v}{v_{B/A}}$$

so that $\qquad\qquad \Delta v = v_{B/A}(\Delta\theta)$

The acceleration of B due to this change in direction is

$$a_B = \frac{\Delta v}{\Delta t} = \frac{v_{B/A}(\Delta\theta)}{\Delta t}$$

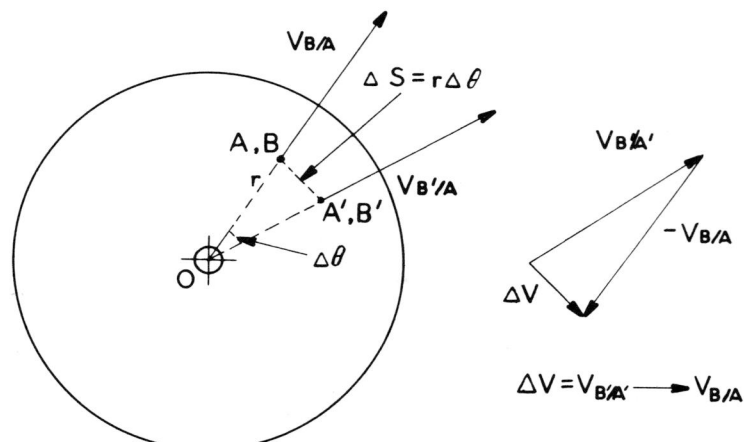

FIG. 3-40 CHANGE IN VELOCITY OF THE BLOCK BE-
CAUSE OF ROTATION OF THE DISK.

But $\Delta\theta/\Delta t = \omega$, therefore $a_B = v_{B/A}\omega$. The direction of this acceleration is perpendicular to OA (tangential) or parallel to Δv and consistent with the sense of ω.

This accounts for the first of the omitted accelerations. Now for
the second one.

As B moves radially outward its radius OB increases to OB'.
Since $v = r\omega$ the tangential velocity $v_{B/A}$ must increase. See Fig. 3-41.
In a very small increment of time Δt the radius OB changes by an
amount Δr, so that the change in $v_{B/A}$ will be $(\Delta r)\omega$. The acceleration

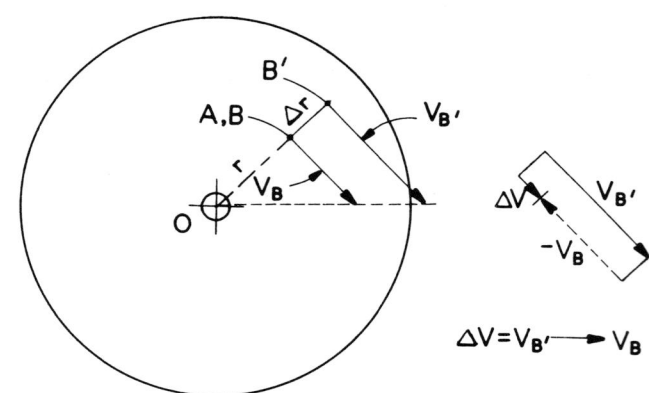

FIG. 3-41 CHANGE IN VELOCITY OF THE BLOCK BE-
CAUSE OF RADIAL MOTION.

of B due to such a change in r will be

$$a_B = \frac{\Delta v}{\Delta t} = \frac{(\Delta r)\omega}{\Delta t} = v_{B/A}\omega$$

Note that $\Delta r/\Delta t$ is a velocity, having units of length per time. This acceleration also is perpendicular to OA, that is, tangential, and consistent with the sense of ω.

The additional two terms for the acceleration of B relative to A are equal and in the same direction. Therefore they may be added to give a single term

$$2v_{B/A}\omega$$

This term is called the Coriolis acceleration, named after its discoverer. The Coriolis vector is always perpendicular to $v_{B/A}$ and is directed as if it had been rotated through 90° about its tail end, with $v_{B/A}$ in the same direction as ω.

The Coriolis acceleration term must be added to the usual acceleration equation to give

$$a_B = a_A \mapsto a_{B/A} \mapsto 2v_{B/A}\omega \qquad \text{(Fig. 3-42)}$$

To recapitulate, A is a point on the disk coincident with B, while B is a point on the sliding block. The term $a_{B/A}$ refers to the acceleration of B in a radial direction along the slot relative to A. It is the acceleration of B when the disk is stationary. The complete acceleration vector diagram is shown in Fig. 3-42, assuming a constant angular velocity for the disk. If however the disk has also angular acceleration, a tangential acceleration for A must be included (Fig. 3-43).

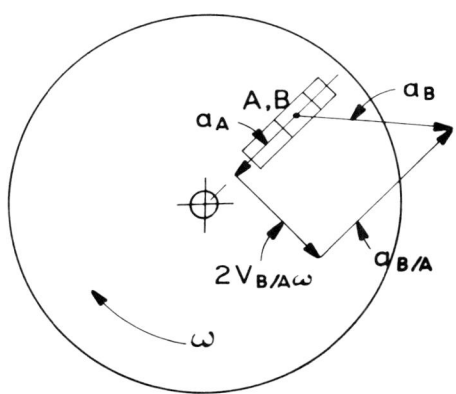

FIG. 3-42   ACCELERATION DIAGRAM FOR BLOCK B.

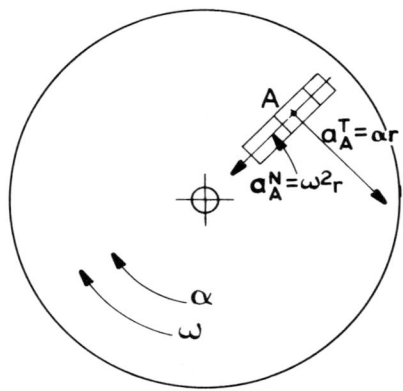

FIG. 3-43   ANGULAR ACCELERATION OF THE DISK.

The point B may have radial movement also; it may move along a curvilinear path on the disk. If such is the case then $a_{B/A}$ has a normal and a tangential component as indicated in Fig. 3-44. The Coriolis component is however unchanged.

Example.   The rotating crank of Fig. 3-45 has a constant angular velocity as shown. The sliding block B is 0.5 meters from the pivot point A, moving radially out at 1.5 m/s. Determine:

a)   The normal, tangential, and Coriolis components of acceleration of the block

b)   The resultant acceleration of the block

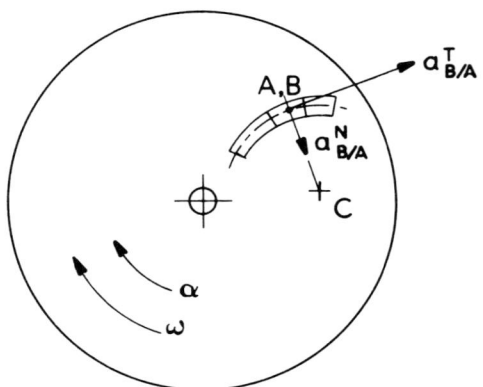

FIG. 3-44   ADDITIONAL ACCELERATION COMPONENTS WHEN THE BLOCK MOVES IN A SLOT WITH CURVATURE.

Solution.

$$a_n = r\omega_c^2 = 0.5(1.5)^2 = 1.125 \text{ m/s}^2$$
$$a_t = r\alpha = 0$$

Coriolis acceleration $= a_{CO} = 2V_B\omega_C = 2(1.5)1.5 = 4.5 \text{ m/s}^2$ in the downward direction.

The resultant acceleration is shown in Fig. 3-45.

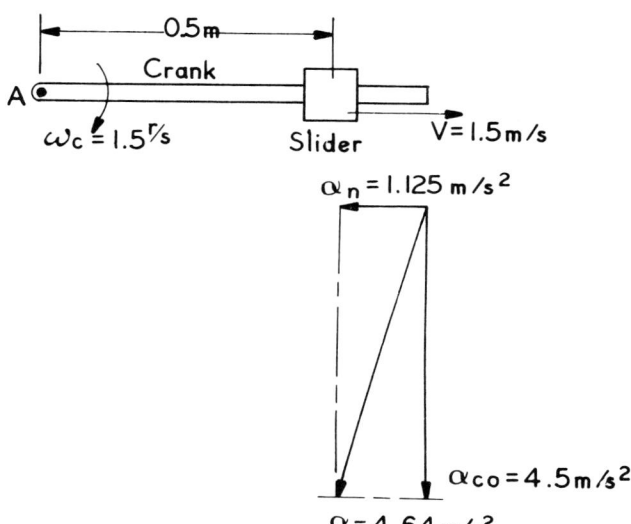

FIG. 3-45   CORIOLIS EXAMPLE.

PROBLEMS

1. A racing quarter horse makes 2.9 strides per second with each leg, its leg oscillating through an angle of 120°. What is the average angular velocity of its leg expressed in radians per second?

2. A point on the circumference of a flywheel is 23 cm from the center of rotation. If the flywheel is rotated at 1500 rpm, what is the tangential velocity of the point?

3. An automobile travels around a highway curve at 55 mph. If the radius of the curve is 2400 ft, what is the angular velocity of the automobile in rad/sec?

4. A flywheel rotates for 20 seconds with a clockwise angular acceleration of 4 rad/sec². Determine its angular displacement
    a) If it starts from rest
    b) If its initial angular velocity is 8 rad/sec clockwise
    c) If its initial angular velocity is 8 rad/sec counterclockwise.

5. An aircraft has been traveling due north and then makes a turn with a radius of 4500 meters to travel due east. The aircraft maintains a constant speed of 700 km/h throughout. Find:
    a) The angular velocity in rad/sec during the turn
    b) The normal acceleration during the turn

6. After World War II the railways converted from steam to diesel engines for a number of reasons, one of them being kinematic. The driving wheels of a steam locomotive must be connected together with heavy side rods (Fig. 1-2), and the normal acceleration of these rods imposed heavy dynamic loads on the track.
    What force is exerted due to normal acceleration of one side rod weighing 500 lb at a radius of 1 ft from the rotational axis if the locomotive is traveling 65 mph on driving wheels 6 ft in diameter?

7. An eccentric cam rotates at 120 rpm clockwise (Fig. P3-7). The cam radius is 35 mm and the eccentricity $e = 10$ mm. Determine:
    a) The normal acceleration of point P
    b) The magnitude and direction of the linear velocity of points P and Q at the position shown in the figure.

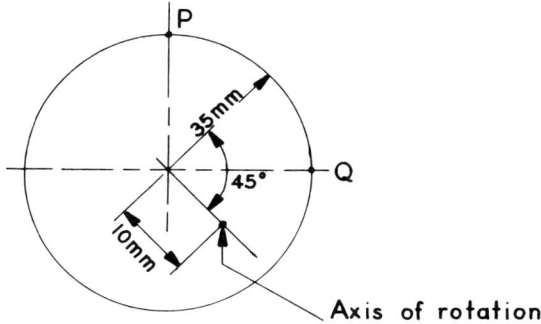

FIG. P3-7 ECCENTRIC CAM.

8. The point *P* lies on the rim of the wheel shown in Fig. P3-8
    The wheel has a clockwise angular velocity of 5 rad/sec and
    a counterclockwise angular acceleration of 20 rad/sec². Find
    the magnitude and direction of the total acceleration of *P*.

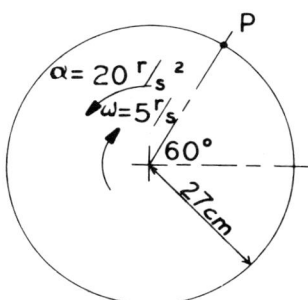

FIG. P3-8 DECELERATING WHEEL.

9. A flywheel is braked so that in 15 seconds it is brought to a
    stop from 1800 rpm with a uniform acceleration rate. What is
    the angular acceleration and the angular displacement during
    deceleration?

10. An engine shaft is accelerated at a uniform rate from 900 rpm
    to 1500 rpm while turning through 100 revolutions. What is
    the angular acceleration and the period of time required to
    change the speed of the shaft?

11. The wheel shown in the accompanying figure rolls on a rail as shown. Angular velocity of the wheel is 3 rad/sec counter-clockwise and angular acceleration is 3 rad/sec² clockwise. Determine the magnitude and acceleration of point *P*.

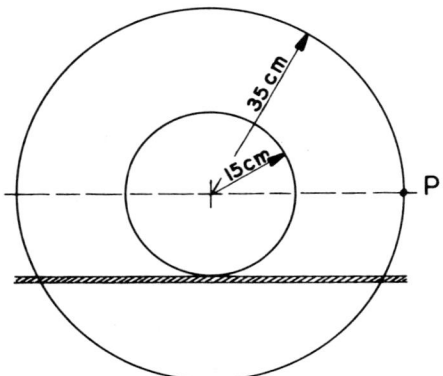

FIG. P3-11   STEPPED WHEEL.

12. Weights *A* and *B* are hung from cables wrapped around the double winch as shown in Fig. P3-12. Weight *A* is moving down, its velocity increasing from 15 cm/s to 25 cm/s in 2 seconds. For this 2-second interval, determine
    a)   The acceleration of *A* and *B*
    b)   The displacements of *A* and *B*
    c)   The angular acceleration of the winch
    d)   The angular displacement of the winch.

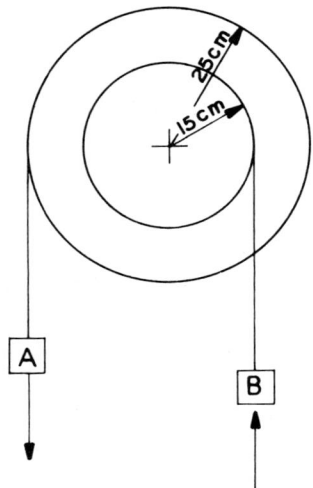

FIG. P3-12   DOUBLE WINCH.

13. For the slider-crank mechanism of Fig. P3-19, find the acceleration of C at TDC using an acceleration polygon. The crank has an angular velocity of 30 rpm.

14. Using the concept of the instant center, explain why Watt's linkage of Fig. 1-16 should give linear displacement of point E when in the position shown.

15. Using the concept of the instant center, explain why the linkage of Fig. 1-17 can produce only a linear displacement of point D.

16. Where is the instant center for link BC in the position shown in Fig. P1-4?

17. Using the instant center method, determine the angular velocity of BC in Fig. 3-5.

18. The crank of the offset slider-crank mechanism shown has an angular velocity of 3 rad/sec counterclockwise and an angular acceleration of 1 rad/sec² clockwise. Determine the acceleration of the slider C.

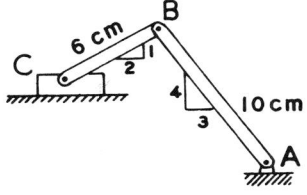

FIG. P3-18   OFFSET SLIDER CRANK.

19. Find the linear velocity of the slider in the slider-crank shown using relative velocity and check by the instant center method. The crank has an angular velocity of 30 rpm.

FIG. P3-19   SLIDER-CRANK.

20. Crank *AB* of the four-bar linkage shown rotates at 12 rad/ sec. Find the angular velocity of *CD* by relative velocity.

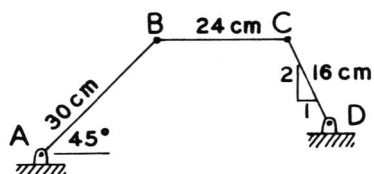

FIG. P3-20   FOUR-BAR LINKAGE.

21. For the linkage of Fig. P3-20, find the angular velocity of *CD* using the instant center method.

22. For the linkage in the position shown, determine the accelera- tion of *C*. Crank *AB* rotates at 40 rad/sec clockwise.

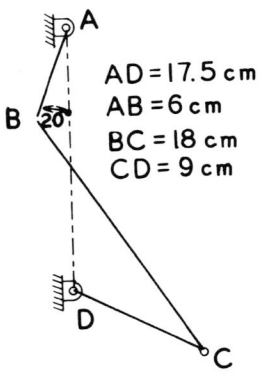

AD = 17.5 cm
AB = 6 cm
BC = 18 cm
CD = 9 cm

FIG. P3-22   FOUR-BAR LINKAGE.

23. In the quick-return mechanism shown, the crank turns at 60 rpm.
   a) Determine the length of the stroke of *E*.

b) Determine the ratio of the time of the advance to the return stroke

c) Determine v_E during the fast and the return stroke.

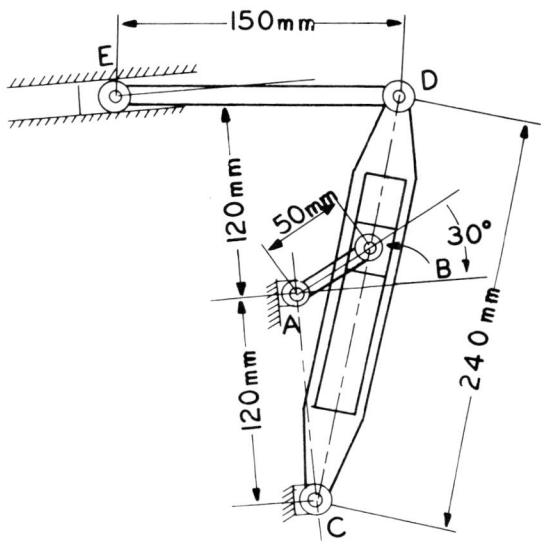

FIG. P3-23   QUICK-RETURN MECHANISM.

24. In the mechanism shown the crank AB rotates at 0.5 rad/sec. Dimensions not shown in the figure are the following:

AB = 45 mm	PF = 30 mm	FP = 60 mm
BP = 120 mm	PE = 120 mm	ED = 84 mm

Determine the linear velocity of F and the angular velocity of EF when the mechanism is in the position shown in the figure.

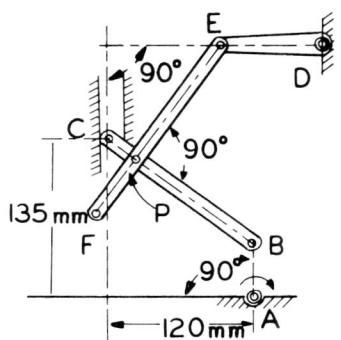

FIG. P3-24   MULTIBAR MECHANISM.

25. The mechanism shown in the figure is driven by crank *AB* at 1 rad/sec counterclockwise. Link *CTD* rotates about a fixed axis *T* and *EF* about a fixed axis *F*. Determine the angular velocity of *DE* and of *EF* when the mechanism is in the position shown.

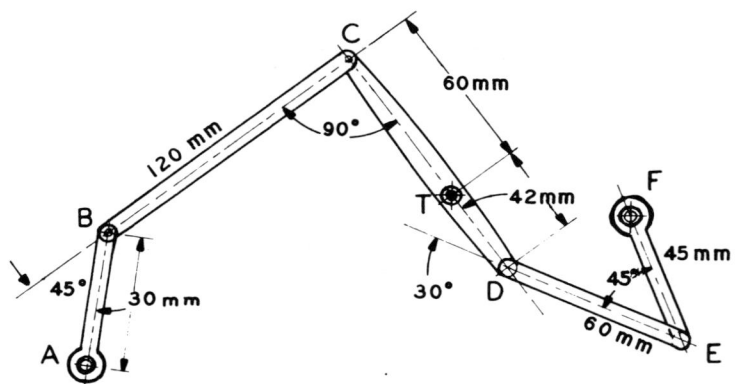

FIG. P3-25   DOUBLE FOUR-BAR LINKAGE.

26. In the figure, crank *AB* rotates at 60 rpm. The dimensions of the links are these:

AB= 30 mm	*DA*= 120 mm
BEC= 135 mm	*EF*= 60 mm
BE= 60 mm	*FG*= 120 mm
CD= 60 mm	*DG*= 150 mm

Determine the linear velocity of point *F* and the angular velocity of the link *EF*.

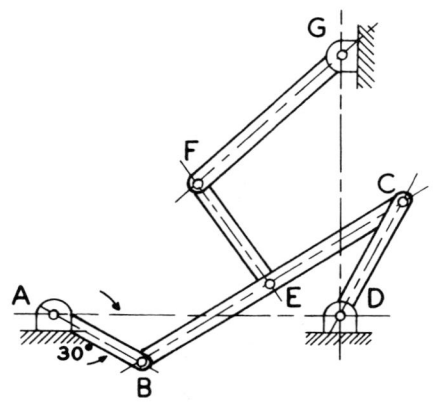

FIG. P3-26   DOUBLE FOUR-BAR LINKAGE.

27. The wheel in the figure has a translational velocity of 8.3 m/s as shown. Find the angular velocity of the wheel and the velocities of points *B* and *C* in both magnitude and direction. The wheel does not slip.

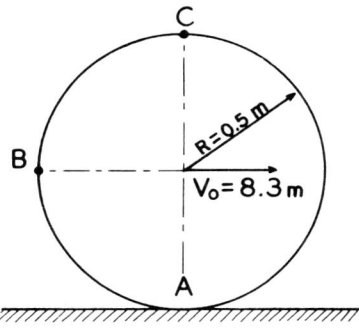

FIG. P3-27   ROLLING WHEEL.

28. The wheel in the figure rolls without slipping at an angular velocity of 4 rad/sec as shown. Find the velocity of point *B* in both magnitude and direction.

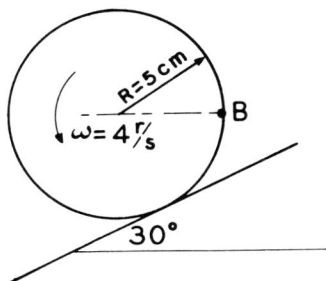

FIG. P3-28   WHEEL ROLLING ON SLOPE.

29. The wheel in the figure rolls without slipping on its hub at an angular velocity of 30 rpm clockwise. For the position shown, determine the velocities of points A and B.

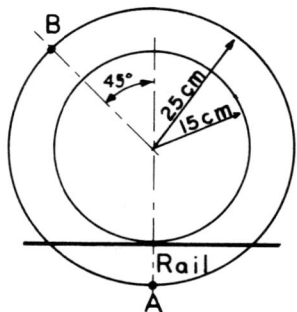

FIG. P3-29   HUBBED WHEEL.

30. If the wheel of problem 29 has an angular acceleration of 0.2 rad/sec^2 clockwise, all other conditions and dimensions remaining as before, what is the total acceleration of points A and B in both magnitude and direction?

31. Link AB is subject to the accelerations shown in the figure. Find the absolute acceleration of point B.

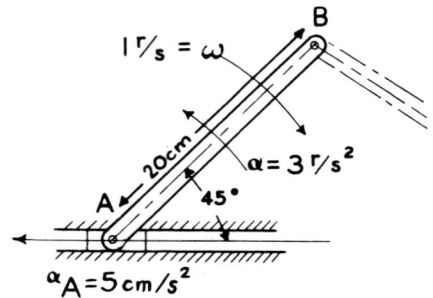

FIG. P3-31   ACCELERATION IN A LINK.

32. In the figure, gear A is rotating at 10 rpm counterclockwise. Gear A has a pitch diameter of 6 cm and gear B of 10 cm. Link CD fastened to gear A has a length of 6 cm and link DE fastened to gear B has a length of 10 cm. Both links pivot at their ends. In the position shown find the velocity of point E.

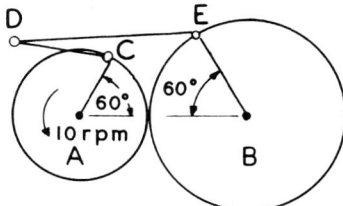

FIG. P3-32 GEARED LINKAGE.

33. The figure shows a cone roller toroidal friction drive, providing an infinitely variable speed ratio between input and output shafts.
    a) explain the principle of operation of this mechanism
    b) if $R_{IN} = 16$ cm and $R_{OUT} = 8$ cm, with the input shaft rotating at 1800 rpm, what is the speed of the output shaft?

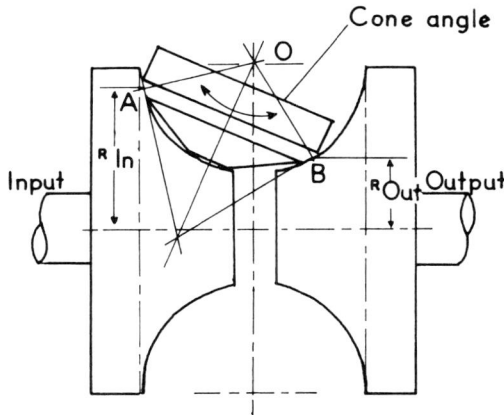

FIG. P3-33 CONE ROLLER TOROIDAL FRICTION DRIVE.

34. On a slippery hill, the drive wheels of an automobile rotate at 4 rps while the vehicle moves down the hill at 1 mps. Tire diameter is 0.65 meters. What is the sliding velocity between tire and road?

35. The figure shows two teeth of a pair of mating gears in contact on the tangent line *TT*. The driving gear has a pitch diameter of 20 cm and rotates at 800 rpm counterclockwise. The driven gear has a pitch diameter of 35 cm. Determine the sliding velocity, assuming that the contact between gears is at the pitch radius.

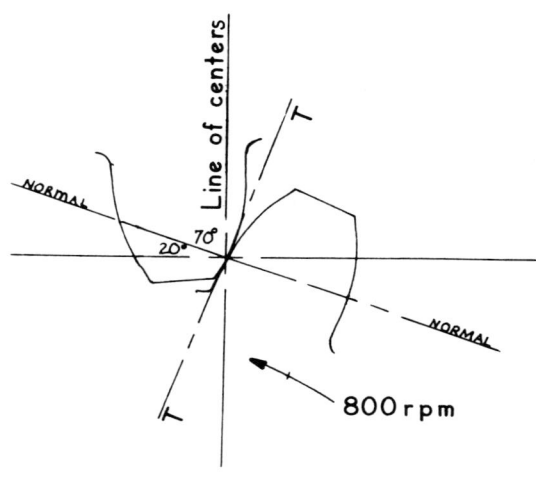

FIG. P3-35

36. In the Geneva intermittent motion mechanism of the figure a pin on the driving wheel engages a slot in the spider as shown, to rotate the spider one-quarter turn for every rotation of the driver. Rotional speed of the driver is 1 rad/sec. Determine the velocity of sliding.

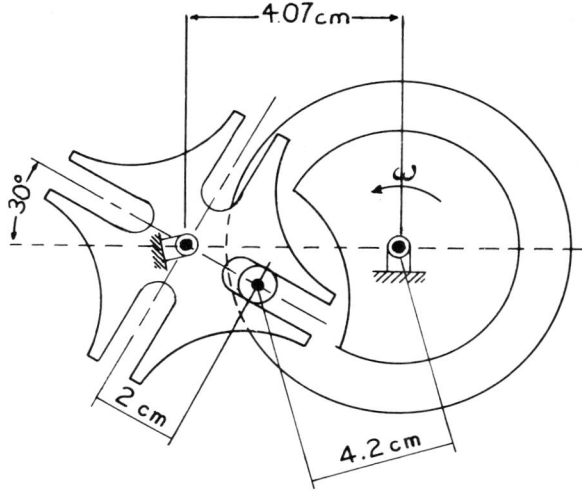

FIG. P3-36 GENEVA MECHANISM.

37. In the figure the cam C rotates clockwise at 1 rps about an axis Q; the follower swings about an axis R. In the position shown, the distance from the axis of rotation of the cam to the point of contact with the follower is 3.8 cm. Angles to the vertical are indicated in the figure. Determine the velocity of sliding.

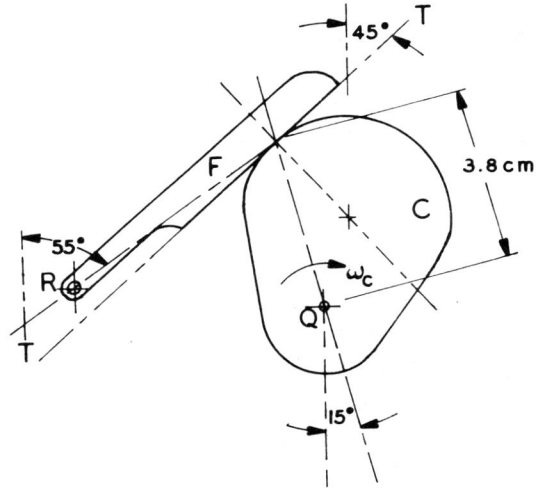

FIG. P3-37 CAM AND FOLLOWER WITH SLIDING CONTACT.

38. In the figure, point *A* is the axis of rotation of the arm *AB* and also the center of curvature of the ring gear *R*. The planetary gear *G* meshes with the ring gear *R*. The point *P* is the point on the planetary gear in contact with the ring gear. *AB*=12 cm and the radius of the ring gear is 18 cm. Plot the path of *P* on the gear *G* as the arm rotates through 90°. What is the name given to such a curve?

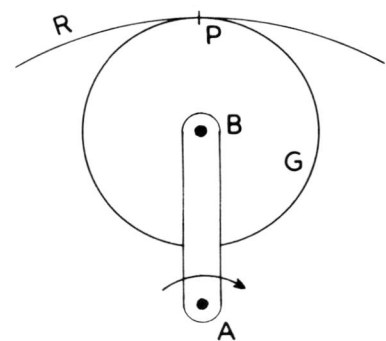

FIG. P3-38   PLANETARY GEAR AND RING GEAR.

39. The point on a rolling wheel in contact with its ground surface has zero velocity. Examine the cycloidal curve of Fig. 3-27. Does the cycloid substantiate this statement?

40. Lay out an involute curve by unwinding through π radians.

41. A flanged wheel such as a railroad wheel, produces a reversal of motion once every revolution. Lay out the curve of displacement of a point on the flange of such a wheel.

42. Set up a formula for the displacement of the follower in a swash-plate mechanism. Include the angle of the swash-plate.

43. Draw the epitrochoidal curve traversed by a point on the circumference of a generating circle which has a radius of one-half that of the fixed circle.

44. For a generating circle with a radius half that of the fixed circle, draw the epitrochoidal curve traveled by a point on the generating circle located at the half-radius from the center.

45. Draw the epitrochoidal curve traversed by a point on the circumference of a generating circle that has a radius one-third that of the fixed circle.

46. A Scotch yoke is driven by a crank 6 cm long rotating at 5 rad/sec. Determine the acceleration of the yoke when the crank is 30° from vertical. The yoke reciprocates vertically.

47. In the figure the crank is vertical, with an angular velocity of 5 rad/sec and an angular acceleration of 1 rad/sec², both counterclockwise. The slider on the crank at this instant is 0.5 meters from the axis of rotation, moving upward at a velocity of 0.3 m/s. Find the normal, tangential, and Coriolis components of acceleration of the slider.

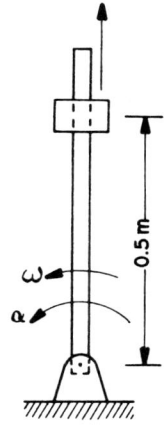

FIG. P3-47   CORIOLIS ACCELERATION.

48. The crank in the figure rotates clockwise at 5 rad/sec. The pin connecting the vertical arm and the slider is designated *B*. Take point *A* to be a point on the crank directly beneath *B*. Find the total acceleration of pin *B* and also the angular acceleration of the arm *AB*.

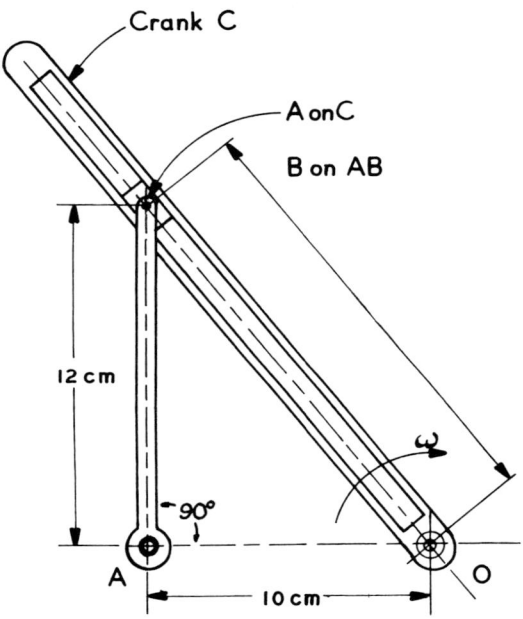

FIG. P3-48 CORIOLIS ACCELERATION IN A CRANK AND SLIDER.

chapter four

Cams

4.1. TYPES OF CAMS.

The use of cams to convert constant angular velocity into complex linear or other velocity cycles is so common that few machines are without cams. Typewriters, sewing machines, copying and printing machines, can openers, and automobiles are only a few of the machines that require cams. A small cam on the distributor shaft of an automobile controls ignition in the cylinders: the camshaft opens and closes intake and exhaust valves.

It is difficult to obtain complex kinds of variable motion by means of pin-jointed linkages, but the task becomes relatively easy if cams are used. The cam, too, is a compact mechanism. An example of a cam-controlled motion of some complexity is shown in Fig. 4-1. The displacement-time relationship of the follower is controlled by the cam profile. The flat length in the middle of the cam profile is called a *dwell*; during a dwell there is no movement of the follower.

The cam of Fig. 4-1 is called a translation cam. Such cams are rarely used. Almost always a cam is attached to a shaft rotating at a constant angular velocity. Some types of flat or disk cams are illustrated in Fig. 4-2, while a cylindrical or drum cam is shown in Fig. 4-3. The roller follower is preferred over point followers and flat-faced followers because of reduced wear of the surfaces of cam and follower.

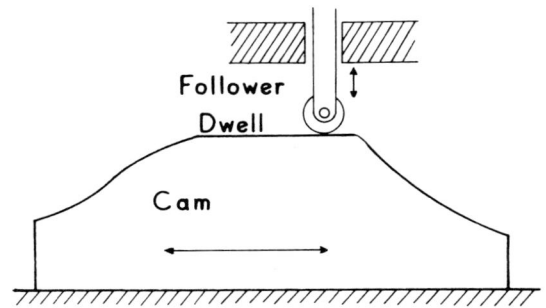

FIG. 4-1   TRANSLATION CAM.

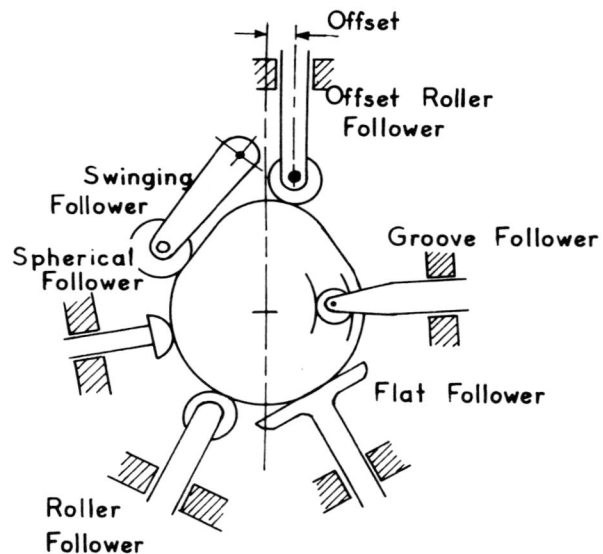

FIG. 4-2   TYPES OF FLAT CAM FOLLOWERS.

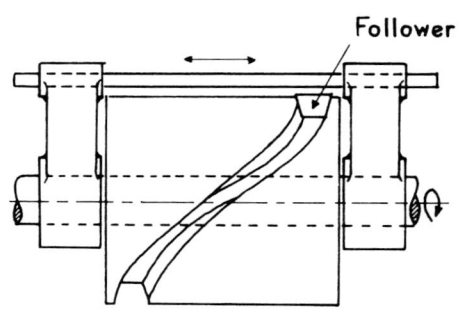

FIG. 4-3   DRUM CAM WITH FOLLOWER MOTION
PARALLEL TO CAMSHAFT.

4.2.   A BASIC CAM DESIGN.

A basic cutoff operation on an automatic lathe provides an example of cam design. In this operation 12-ft lengths of steel bar stock 0.500″ (inches) in diameter are to be cut into lengths $1\frac{1}{4}$ in. long using a cutoff tool (parting tool). The parting tool, when in the retracted position, is positioned 0.010 in. away from the workpiece, and must be advanced into the workpiece slightly past its center in order to part off the piece. See Fig. 4-4. The total movement of the tool is taken as 0.270 inches. This movement severs the piece, after which the parting tool must be rapidly retracted. The tool remains in the retracted position while the bar is (automatically) moved ahead to a fixed stop to expose another $1\frac{1}{4}$ in. length for cutoff.

The cutoff tool must be fed into the workpiece 0.002 in. per revolution of the workpiece, and the speed of the lathe spindle is taken at 800 rpm. Therefore, the number of revolutions for cutoff = 0.270/0.002 = 135 revolutions. Another 25 revolutions are allowed for feeding the bar forward for the next cutoff cycle. Therefore total revolutions to produce one piece are 160. A cam must be designed to produce this cycle of motion in the cutoff tool. The movements of the cam follower must be the movements of the tool.

In designing cams for automatic lathes it is common to divide the circumference of the cam into 100 divisions rather than 360 degree divisions, and cam blanks may be marked for 100 divisions. One hundredth (100th) therefore equals 3.6°. In the present case 160 revolutions of the lathe spindle complete one machining cycle. Also, one rotation of the cam (100 100ths) must correspond to one machining cycle. Therefore, there must be 1.6 spindle revolutions per cam 100th. Cutting off requires 135 rev/1.6 rev per 100th or 84 100ths.

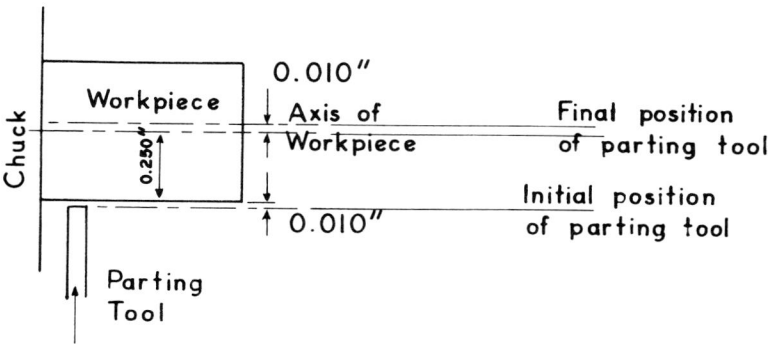

FIG. 4-4   A CUTOFF OPERATION ON A LATHE.

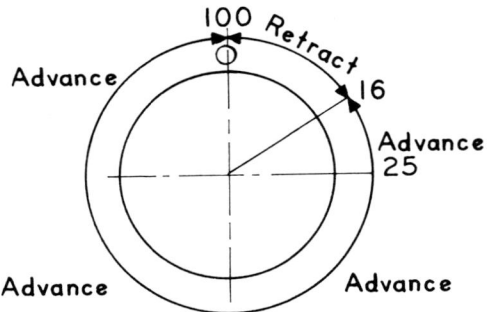

FIG. 4-5 BASIC LAYOUT FOR CUTOFF CAM USING CAM
        HUNDREDTHS.

Feeding the bar stock requires 25 rev/1.6 or 15.6 100ths, or, say, 16
100ths. In these 16 100ths the tool must be quickly retracted. The
basic layout for the cam profile is given in Fig. 4-5.

The cam is profiled as follows. Suppose the base circle of the
cam (the circle defined by the smallest radius of the cam profile) has
a diameter of 1.500 in. The cam must lift the follower a distance of
0.270 in. in 84 100ths. Therefore at the end of the cut the cam radius
must measure

$$\frac{1.500}{2} + 0.270 = 1.02''$$

The contour of the cam must in 84 hundredths lift the follower 0.270
in. or 0.0032 in. per 100th. Since the radial distance of the contour
increases 0.0032 in./100th, the cam profile is laid out as shown in Fig.
4-6 for the tool advance. Suppose the maximum cam diameter is
placed at the zero 100th. Then the final cam profile is as in Fig. 4-7,
allowing 2 100ths for retraction of the tool. We are assuming that

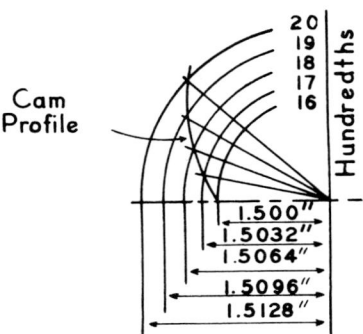

FIG. 4-6 CAM LAYOUT FOR TOOL ADVANCE OF
        0.0032''/100TH.

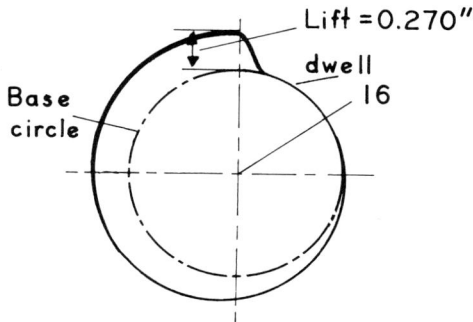

FIG. 4-7   FINAL CUTOFF CAM.

the follower can maintain contact with the cam during this rapid retraction.

This cam has been designed to the requirements of the machining cycle. No attention has been paid to considerations of kinematics, and this must next be considered.

Fig. 4-8 is a linear layout of the cam, displacement against 100ths. The velocity of advance and retract is constant. Velocity is the slope of the displacement line. At two points on the cam the velocity changes instantly from zero or to zero; at a third point between advance and retract, the velocity changes instantly from positive to negative. Such instantaneous changes of velocity can be made only with infinite acceleration, and an infinite acceleration can be produced only with an infinitely large force. This is impossible of course; what happens is that the cam and follower deform under the high acceleration forces and reduce the acceleration to some finite value.

Kinematically this is a poor cam design. The sudden changes in velocity can be reduced by changing the follower velocity over a finite period of time. This example illustrates the fundamental problem in the design of cams. What may be a good cam for machine motion may be unacceptable kinematically because of severe acceleration.

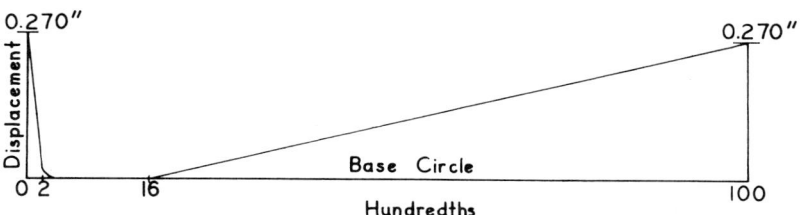

FIG. 4-8   LINEAR LAYOUT OF CAM.

4.3. BASIC FOLLOWER MOVEMENTS.

It is convenient to lay out follower movements as is done in Fig. 4-8. Follower lift is plotted vertically and cam angle horizontally. The zero lift position lies on a circle called the base circle of the cam, that is, all lifts are measured from the base circle.

Parabolic Motion.

It has been explained that a constant velocity cam is not a desirable profile because of uncontrolled accelerations at the points where the velocity changes. A better cam movement is parabolic motion, because it has constant acceleration. The equation $s = \frac{1}{2}at^2$ is a parabola.

The displacement diagram for a parabolic or constant acceleration cam is given in Fig. 4-9. An even number of angle positions is selected for the abscissa. If 8 angle positions are used, then the corresponding lift positions are segments in the proportions $1:3:5:7:7:5:3:1$. For 12 positions the proportions are $1:3:5:7:9:11:11:9:7:5:3:1$. These relative lengths are laid out on any suitable line AB and projected by similar triangles to the vertical lift line as shown in the figure. In the figure the sum of $1+3+5+7+7+5+3+1$ is 32, therefore any convenient line at any angle is drawn 32 units long, such as 32/16 in. or $32 \times 3/16$ in. long. These lift positions are transferred horizontally to produce the cam angle lift curve. The cam follower is lifted and returned at constant acceleration, except at the point where the velocity changes from

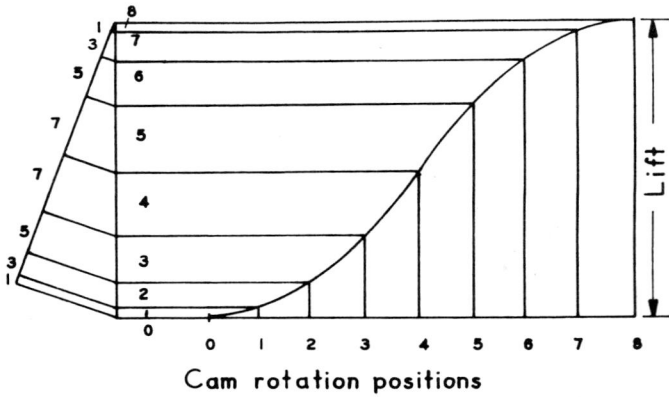

Cam rotation positions

FIG. 4-9   CONSTANT ACCELERATION CAM ACTION.

increasing to decreasing. At this point the acceleration is infinite. Because of this high acceleration, high-speed cams do not use parabolic motion.

Harmonic Motion.

To construct a cam for harmonic motion, use the semicircle as in Fig. 4-10. The cam angles are numbered on both the lift graph and the semicircle. The points on the semicircle are transferred horizontally to the lift graph. The graph shows a follower rise; a follower fall would have the same construction.

Though the acceleration is always finite for harmonic motion, the time rate of acceleration, called jerk, is infinite at the beginning and end of harmonic motion. This type of motion is not recommended for high speed cams.

Cycloidal Motion.

Cycloidal motion provides zero acceleration at the beginning and the end of the follower lift or fall. This type of motion was discussed in Sec. 3-11. In Fig. 4-11, the lift is the circumference of the circle drawn at the base of the diagram. The circle is divided into as many segments as are used on the follower lift layout (12 divisions in the figure). The circle is rolled vertically upward with its center stopped opposite each point on the lift line, and a partial circle is drawn to intersect a vertical from the base circle as shown.

Cycloidal motion has suitably low acceleration for use in high-speed cams.

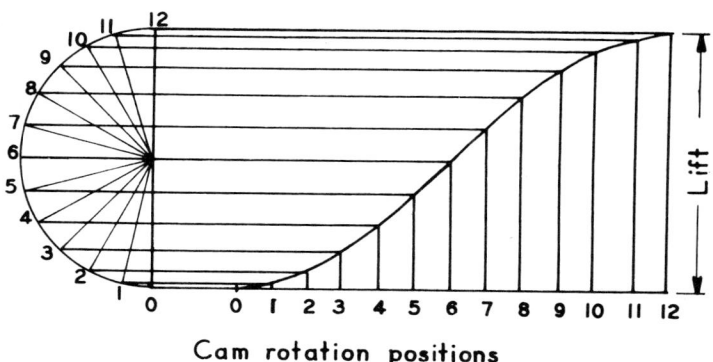

Cam rotation positions

FIG. 4-10   HARMONIC MOTION.

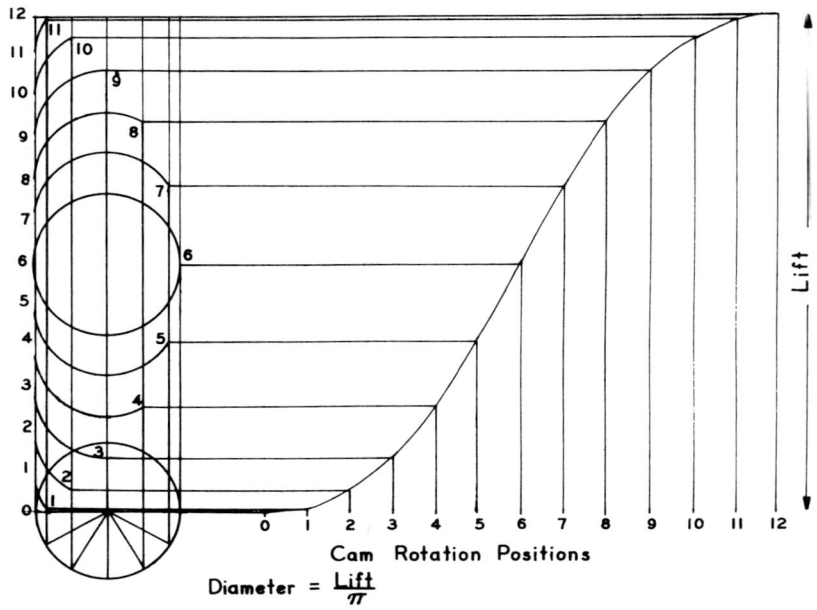

FIG. 4-11 CYCLOIDAL CAM MOTION.

Example. A cam profile is to provide follower motion as follows:

1. Base circle 6 cm in diameter

2. Counterclockwise cam rotation

3. Point follower, with axis of follower on the center of rotation of the cam

4. Follower motion:
   3 cm rise with cycloidal motion in 120° of cam rotation
   dwell for 30°
   harmonic motion for 180° of fall
   dwell for 30°

The cam design is given in Fig. 4-12. Cycloidal lift motion is laid out on the left-hand side; harmonic fall on the right-hand side. One complete revolution of the cycloidal base circle is required for the 120° cycloidal follower lift.

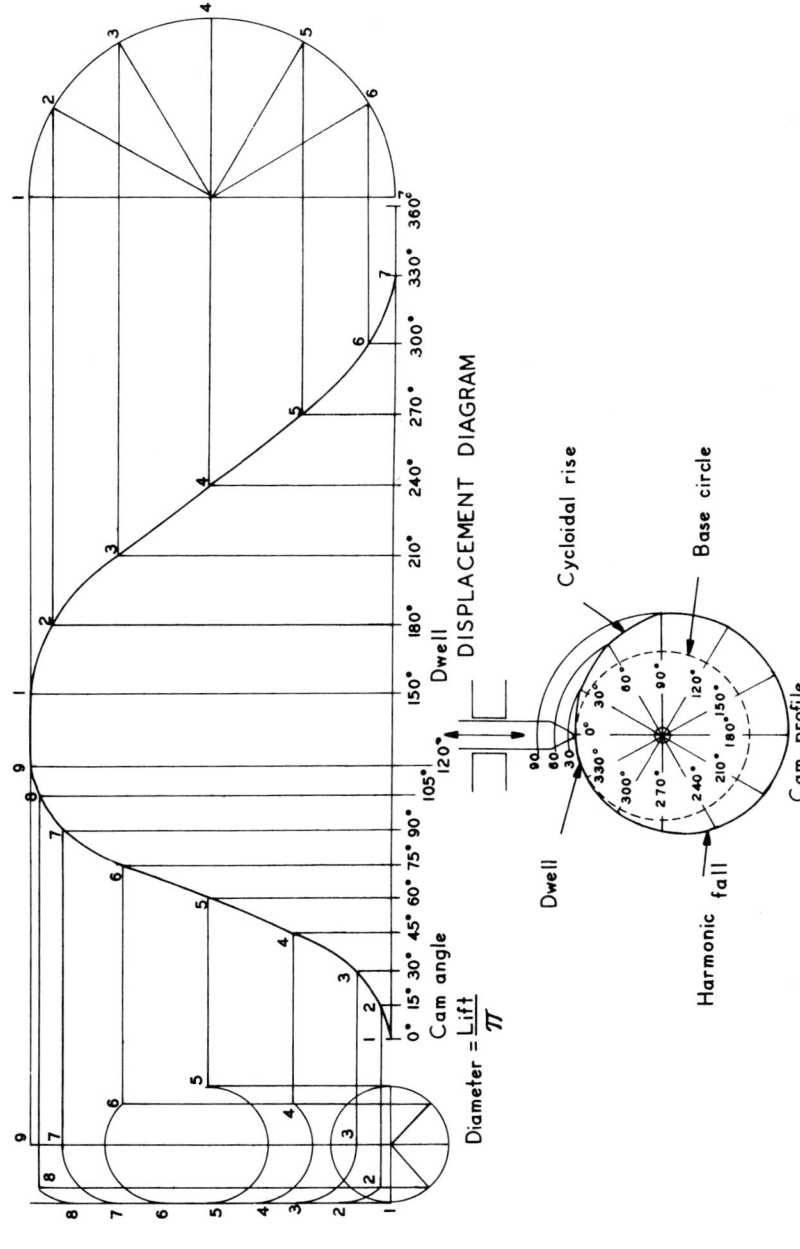

DISPLACEMENT DIAGRAM

Diameter = $\dfrac{\text{Lift}}{\pi}$

Cam angle

0° 15° 30° 45° 60° 75° 90° 105° 120° 150° 180° 210° 240° 270° 300° 330° 360°

Dwell

Cycloidal rise

Base circle

Cam profile

Dwell

Harmonic fall

FIG. 4-12 CYCLOIDAL RISE AND HARMONIC FALL.

Modified Trapezoidal Motion.

This cam motion is a compromise, combining the low peak acceleration of parabolic motion and the continuous change of acceleration of cycloidal motion (see Fig. 4-13). It is named from the shape of the acceleration curve, which resembles a trapezoid with rounded corners. Modified trapezoidal motion is excellent for high-speed cams because of its controlled acceleration. Fig. 4-13 shows the shape of the displacement curve for any lift h and any cam lift angle β. A displacement layout is also shown for 90° of lift.

A cam to provide this motion is most conveniently designed from a table of constants. Such a table is supplied in Table 4-4. The use of these constants is explained below.

In Fig. 4-14 a comparison of the dynamic characteristics of the three types of motion used in high-speed cams is made. All three are given in terms of the same lift h over the same cam angle. The cycloidal and modified trapezoidal curves have the same maximum velocity, which is higher than that for the harmonic curve. The

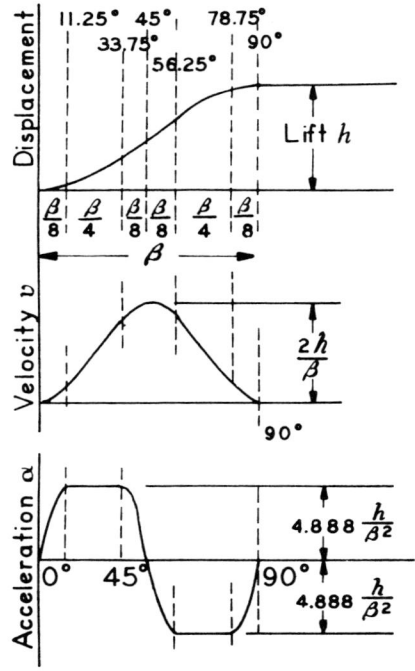

FIG. 4-13   MODIFIED TRAPEZOIDAL CAM MOTION.

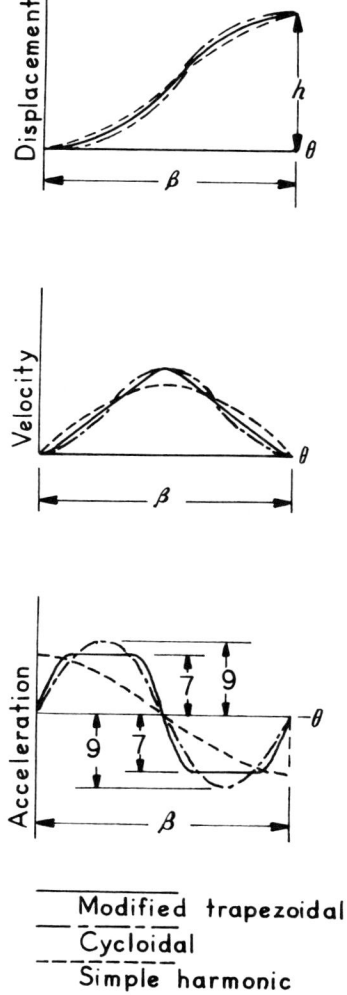

FIG. 4-14 COMPARISON OF HIGH-SPEED CAM MOTIONS.

harmonic and modified trapezoidal curves give the same maximum acceleration, while the maximum acceleration of the cycloidal curve is 9/7 as large as these.

By comparison, the constant-velocity curve of Fig. 4-15 is not suited to high-speed cam action. Accelerations are infinite (or in the practical case, finite, but severe) and result in very high forces.

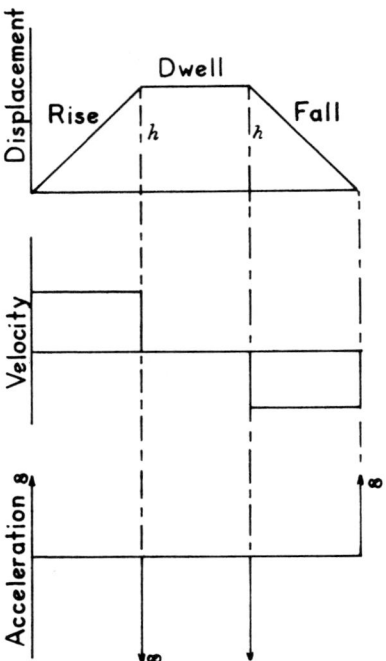

FIG. 4-15   DYNAMIC CHARACTERISTICS OF A CONSTANT
VELOCITY CAM.

4.4.   CAM TABLES.

In the example of a cam for a parting tool, a precision of a few
"tenths" (ten-thousandths of an inch) was implied. Such precision is,
of course, beyond the capacity of graphical technique. For accurate
cam layouts, recourse must be made to cam tables. Table 4-1 gives
displacements for uniform acceleration; Table 4-2 for harmonic motion; Table 4-3 for cycloidal motion; and Table 4-4 for modified
trapezoidal motion. In each table the amplitude is divided into 120
angular increments. The use of these tables is illustrated by the
following situations. Suppose harmonic motion is selected and a total
follower rise of 1.000 in. is required in 60° of cam rotation. The table
gives a 1.000 in. rise in 120 degrees. So for an actual 60° of cam
rotation, 2° in the table represents 1° in the actual cam. Hence for

TABLE 4-1 CAM DISPLACEMENTS
FOR UNIFORM ACCELERATION

θ	y	θ	y	θ	y
1	0.000139	41	0.233471	81	0.788745
2	0.000556	42	0.244999	82	0.799439
3	0.001250	43	0.256804	83	0.809856
4	0.002222	44	0.268887	84	0.819995
5	0.003472	45	0.281248	85	0.829856
6	0.005000	46	0.293887	86	0.839439
7	0.006806	47	0.306804	87	0.848744
8	0.008889	48	0.319998	88	0.857772
9	0.011250	49	0.333470	89	0.866522
10	0.013889	50	0.347220	90	0.874994
11	0.016806	51	0.361248	91	0.883188
12	0.020000	52	0.375553	92	0.891105
13	0.023472	53	0.390136	93	0.898744
14	0.027222	54	0.404997	94	0.906105
15	0.031250	55	0.420136	95	0.913188
16	0.035556	56	0.435553	96	0.919994
17	0.040139	57	0.451247	97	0.926522
18	0.045000	58	0.467219	98	0.932772
19	0.050139	59	0.483469	99	0.938744
20	0.055556	60	0.500000	100	0.944438
21	0.061250	61	0.516525	101	0.949855
22	0.067222	62	0.532776	102	0.954994
23	0.073472	63	0.548747	103	0.959855
24	0.080000	64	0.564441	104	0.964438
25	0.086806	65	0.579858	105	0.968744
26	0.093889	66	0.594997	106	0.972772
27	0.101250	67	0.609858	107	0.976522
28	0.108889	68	0.624441	108	0.979994
29	0.116806	69	0.638746	109	0.983188
30	0.125000	70	0.652774	110	0.986105
31	0.133472	71	0.666524	111	0.988744
32	0.142222	72	0.679996	112	0.991105
33	0.151250	73	0.693190	113	0.993188
34	0.160555	74	0.706107	114	0.994994
35	0.170138	75	0.718746	115	0.996522
36	0.180000	76	0.731107	116	0.997772
37	0.190138	77	0.743190	117	0.998744
38	0.200555	78	0.754995	118	0.999438
39	0.211249	79	0.766523	119	0.999855
40	0.222221	80	0.777773	120	1.000000

TABLE 4-2   CAM DISPLACEMENTS
FOR HARMONIC MOTION

θ	y	θ	y	θ	y
1	0.000171	41	0.261420	81	0.761249
2	0.000685	42	0.273004	82	0.772319
3	0.001541	43	0.284744	83	0.783203
4	0.002739	44	0.296631	84	0.793892
5	0.004277	45	0.308658	85	0.804380
6	0.006155	46	0.320816	86	0.814660
7	0.008372	47	0.333096	87	0.824724
8	0.010926	48	0.345491	88	0.834565
9	0.013815	49	0.357992	89	0.844177
10	0.017037	50	0.370590	90	0.853553
11	0.020590	51	0.383277	91	0.862687
12	0.024471	52	0.396044	92	0.871572
13	0.028679	53	0.408882	93	0.880202
14	0.033209	54	0.421782	94	0.888572
15	0.038060	55	0.434736	95	0.896676
16	0.043227	56	0.447735	96	0.904508
17	0.048707	57	0.460770	97	0.912063
18	0.054496	58	0.473832	98	0.919335
19	0.060591	59	0.486911	99	0.926320
20	0.066987	60	0.500000	100	0.933012
21	0.073679	61	0.513088	101	0.939408
22	0.080664	62	0.526167	102	0.945503
23	0.087936	63	0.539229	103	0.951292
24	0.095491	64	0.552264	104	0.956772
25	0.103323	65	0.565263	105	0.961939
26	0.111427	66	0.578217	106	0.966790
27	0.119797	67	0.591117	107	0.971320
28	0.128427	68	0.603955	108	0.975528
29	0.137312	69	0.616722	109	0.979409
30	0.146446	70	0.629409	110	0.982962
31	0.155822	71	0.642007	111	0.986184
32	0.165434	72	0.654508	112	0.989073
33	0.175275	73	0.666903	113	0.991627
34	0.185339	74	0.679183	114	0.993844
35	0.195619	75	0.691341	115	0.995722
36	0.206107	76	0.703368	116	0.997260
37	0.216796	77	0.715255	117	0.998458
38	0.227680	78	0.726995	118	0.999314
39	0.238750	79	0.738579	119	0.999828
40	0.250000	80	0.750000	120	1.000000

TABLE 4-3   CAM DISPLACEMENTS
FOR CYCLOIDAL MOTION

θ	y	θ	y	θ	y
1	0.000003	41	0.208188	81	0.816808
2	0.000030	42	0.221240	82	0.828728
3	0.000102	43	0.234646	83	0.840250
4	0.000243	44	0.248391	84	0.851365
5	0.000474	45	0.262460	85	0.862065
6	0.000818	46	0.276837	86	0.872343
7	0.001297	47	0.291507	87	0.882195
8	0.001932	48	0.306451	88	0.891616
9	0.002745	49	0.321651	89	0.900603
10	0.003755	50	0.337089	90	0.909154
11	0.004984	51	0.352745	91	0.917270
12	0.006451	52	0.368599	92	0.924949
13	0.008173	53	0.384630	93	0.932195
14	0.010171	54	0.400818	94	0.939010
15	0.012460	55	0.417141	95	0.945398
16	0.015058	56	0.433576	96	0.951365
17	0.017980	57	0.450102	97	0.956917
18	0.021240	58	0.466697	98	0.962061
19	0.024854	59	0.483337	99	0.966808
20	0.028834	60	0.500000	100	0.971165
21	0.033191	61	0.516662	101	0.975145
22	0.037938	62	0.533302	102	0.978759
23	0.043082	63	0.549897	103	0.982019
24	0.048634	64	0.566423	104	0.984941
25	0.054601	65	0.582859	105	0.987539
26	0.060989	66	0.599181	106	0.989828
27	0.067804	67	0.615369	107	0.991826
28	0.075050	68	0.631400	108	0.993548
29	0.082729	69	0.647254	109	0.995015
30	0.090845	70	0.662910	110	0.996244
31	0.099396	71	0.678348	111	0.997254
32	0.108383	72	0.693548	112	0.998067
33	0.117804	73	0.708492	113	0.998702
34	0.127656	74	0.723162	114	0.999181
35	0.137934	75	0.737539	115	0.999525
36	0.148634	76	0.751608	116	0.999756
37	0.159749	77	0.765353	117	0.999897
38	0.171271	78	0.778759	118	0.999969
39	0.183191	79	0.791811	119	0.999996
40	0.195501	80	0.804498	120	1.000000

TABLE 4-4   CAM DISPLACEMENTS FOR MODIFIED
TRAPEZOIDAL MOTION

θ	y	θ	y	θ	y
0	0.00000000				
1	0.00000601	41	0.21668371	81	0.80677140
2	0.00004747	42	0.22892076	82	0.81799010
3	0.00015945	43	0.24149726	83	0.82886930
4	0.00037622	44	0.25441321	84	0.83940910
5	0.00073093	45	0.26766860	85	0.84960940
6	0.00125521	46	0.28126338	86	0.85947030
7	0.00197882	47	0.29519539	87	0.86899170
8	0.00292936	48	0.30945912	88	0.87817370
9	0.00413193	49	0.32404539	89	0.88701620
10	0.00560886	50	0.33894150	90	0.89551920
11	0.00737950	51	0.35413131	91	0.90368280
12	0.00945994	52	0.36959548	92	0.91150700
13	0.01186292	53	0.38531170	93	0.91899170
14	0.01459763	54	0.40125484	94	0.92613680
15	0.01766961	55	0.41739731	95	0.93294270
16	0.02107986	56	0.43370938	96	0.93940900
17	0.02483058	57	0.45015938	97	0.94553580
18	0.02892075	58	0.46671417	98	0.95132330
19	0.03335037	59	0.48333948	99	0.95677130
20	0.03811945	60	0.50000000	100	0.96187980
21	0.04322798	61	0.51666074	101	0.96664890
22	0.04867596	62	0.53328606	102	0.97107840
23	0.05446340	63	0.54984085	103	0.97516860
24	0.06059029	64	0.56629083	104	0.97891940
25	0.06705663	65	0.58260288	105	0.98233060
26	0.07386243	66	0.59874535	106	0.98540319
27	0.08100767	67	0.61468847	107	0.98813786
28	0.08849237	68	0.63040467	108	0.99054079
29	0.09631653	69	0.64586883	109	0.99262118
30	0.10448014	70	0.66105861	110	0.99439175
31	0.11298319	71	0.67595470	111	0.99586860
32	0.12182571	72	0.69054095	112	0.99707109
33	0.13100767	73	0.70480465	113	0.99802155
34	0.14052910	74	0.71873664	114	0.99874507
35	0.15038997	75	0.73233130	115	0.99926925
36	0.16059029	76	0.74558630	116	0.99962386
37	0.17113006	77	0.75850220	117	0.99984053
38	0.18200930	78	0.77107870	118	0.99995239
39	0.19322799	79	0.78331570	119	0.99999376
40	0.20478612	80	0.79521330	120	1.00000000

15, 30, 45, and 60 degrees of cam rotation, the following displacements are used:

Cam Angle	Angle in Table	Follower Displacement
15°	30°	0.146446
30°	60°	0.500000
45°	90°	0.853553
60°	120°	1.000000

If the required cam movement is 0.500 in. instead of 1.000 in., then the tabulated displacements are multiplied by 0.500/1.000 to obtain actual displacements.

Cam Angle	Angle in Table	Follower Displacement
15°	30°	0.073
30°	60°	0.250
45°	90°	0.427
60°	120°	0.500

The tabulated values apply in the same manner to a follower fall.

4.5.   TYPES OF FOLLOWERS

Flat-Face Follower.

First draw the flat follower at its lowest position, that is, tangent to the base circle. At each radial cam station line, draw normals to these tangents at the required displacement, as in Fig. 4-16. These normals represent the face of the follower. Note that the point of contact between cam and follower will shift as the cam rotates. The follower must be wide enough so that the follower can be tangent to the cam at every station, that is, the cam must not go over the edge of the follower. The cam profile is drawn tangent to the follower lines (normal lines). The layout is actually made as though the cam were stationary and the follower rotated about it. The graphical method is shown in the following example.

Example.   A cam must lift a flat-face follower 0.500 in. over 180° in cycloidal motion, dwell for 60°, then provide a cycloidal fall in 60° with a 60° dwell. Design the cam using a 1.750 in. base circle and counterclockwise rotation.

Solution.   For the 180 degree rise, cam stations at 30° will be used. The values for every 20° in Table 4-3 must be used (180°/30° = 120°/20° = 6 equal divisions).

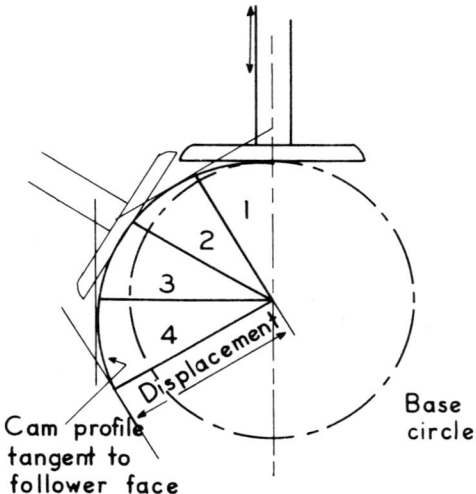

FIG. 4-16 METHOD OF CAM LAYOUT FOR A FLAT-FACE
FOLLOWER.

Cam Angle	Angle in Table	Tabular y	Actual Lift
30°	20°	0.028834	0.014417
60°	40°	0.195501	0.097750
90°	60°	0.500000	0.250000
120°	80°	0.804498	0.402249
150°	100°	0.971165	0.485582
180°	120°	1.000000	0.500000

For the follower fall:

Cam Angle	Angle in Table	Tabular y	Actual Fall
10°	20°	0.0288	0.0144
20°	40°	0.1955	0.0977
30°	60°	0.5000	0.2500
40°	80°	0.8045	0.4022
50°	100°	0.9711	0.4856
60°	120°	1.0000	0.5000

The radial distances for the flat follower are the actual displacement plus base circle radius. These are laid out radially at the appropriate angle, as in Fig. 4-17. The cam profile is then sketched tangent to the follower face. In the figure, the cam profile is drawn from 0° to 240° but not beyond 240°. Between 240 and 270 degrees the follower fall is too steep for a flat follower. An unacceptably

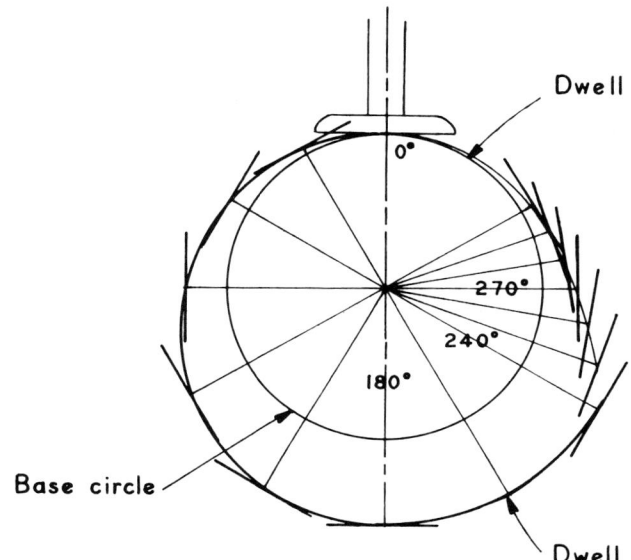

FIG. 4-17 FLAT-FACE FOLLOWER, CLOCKWISE ROTA-
TION. THE FOLLOWER DISPLACEMENTS ARE
EXAGGERATED FOR THE SAKE OF CLARITY.

large cam would solve this problem, but the designer will probably
have to use a roller follower. It is always possible to design an
impractical cam.

Roller Follower.

In laying out a cam for a roller follower it is more convenient to
use the *prime circle* rather than the base circle. The prime circle has
a radius equal to the base circle radius plus the radius of the follower
roller. All lifts are increments above the prime circle. (See the
construction of Fig. 4-18.)

Offset Roller Follower.

The center of rotation of the follower roller may be offset from
the center of rotation of the cam. The offset may be required
because of problems of interference with other parts of the machine,
or to reduce side thrust on the follower. When the offset is to the

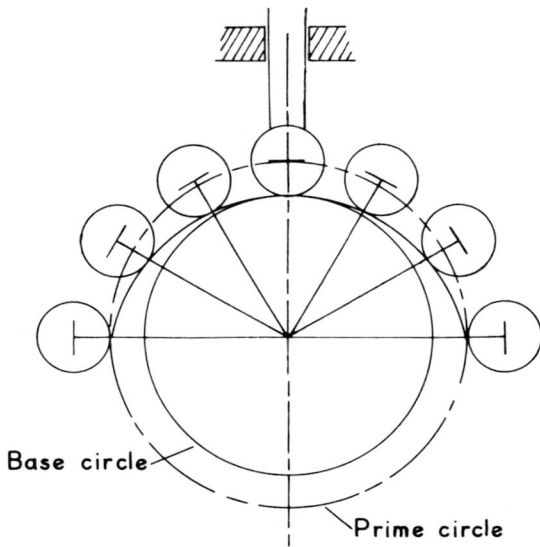

FIG. 4-18   ROLLER FOLLOWER.

right, the cam should rotate counterclockwise, and the cam should use a clockwise rotation if the offset is to the left. With these arrangements side thrust is reduced when the follower rises.

To lay out this type of cam, an offset circle is first drawn; this circle has a radius equal to the offset. Radial lines from the circumference of this circle are drawn at each cam station, as shown in Fig. 4-19. These tangent lines represent the axis of the follower. The displacement of the follower from the base circle is laid out on these tangent lines. The construction is made as though the follower were rotated around the cam. A smooth curve, tangent to the roller position at each station, gives the cam profile.

Pivoted Follower.

To lay out a cam with a pivoted follower as in Fig. 4-20, the base circle is drawn as is a circle with its center also at the axis of rotation of the cam but passing through the pivot axis of the cam. Stations are marked on both circles, as shown in the figure. As the follower lifts, it must move along an arc with a radius given by the length of the oscillating arm. The increments of lift are laid out along this arc at each station and the follower roller, or a partial arc used to represent the roller, is drawn. The cam profile is drawn

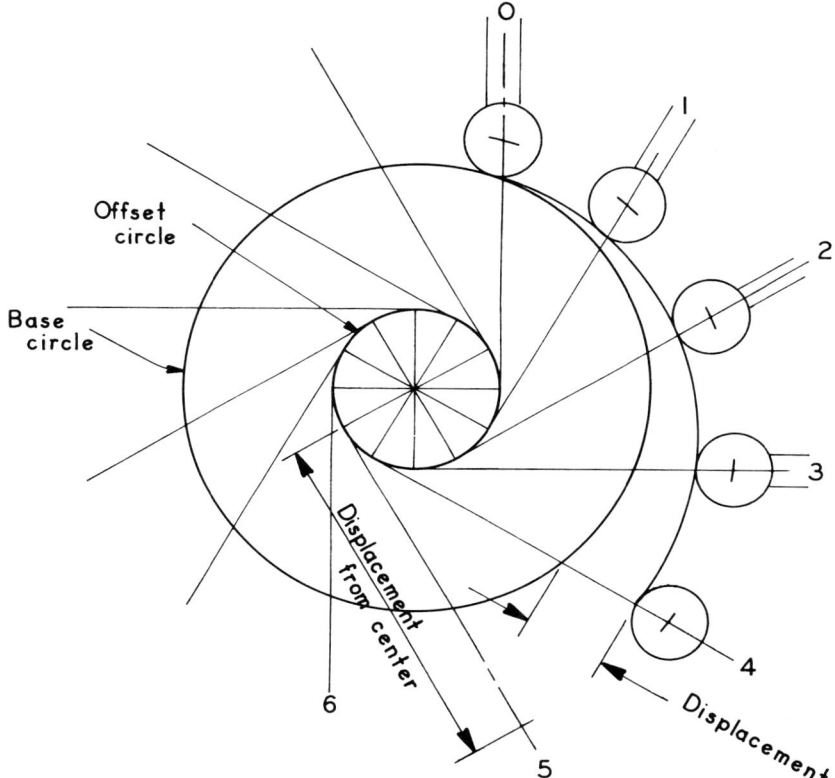

FIG. 4-19   CAM WITH OFFSET FOLLOWER, COUNTER-
          CLOCKWISE ROTATION.

tangent to these roller arcs. As usual, the construction is based on
rotating the follower around the cam.

If the oscillating follower has a flat face the construction is
similar. As with any flat-face follower, the point of contact of the
follower with the cam shifts. The flat face must be made long
enough that it extends beyond the farthest point of contact with the
cam.

4.6.   PRESSURE ANGLE.

The pressure angle, or jamming angle, is the angle between the
axis of the follower and the normal line at the contact point between
follower and cam. This is angle α in Fig. 4-21, which shows the forces

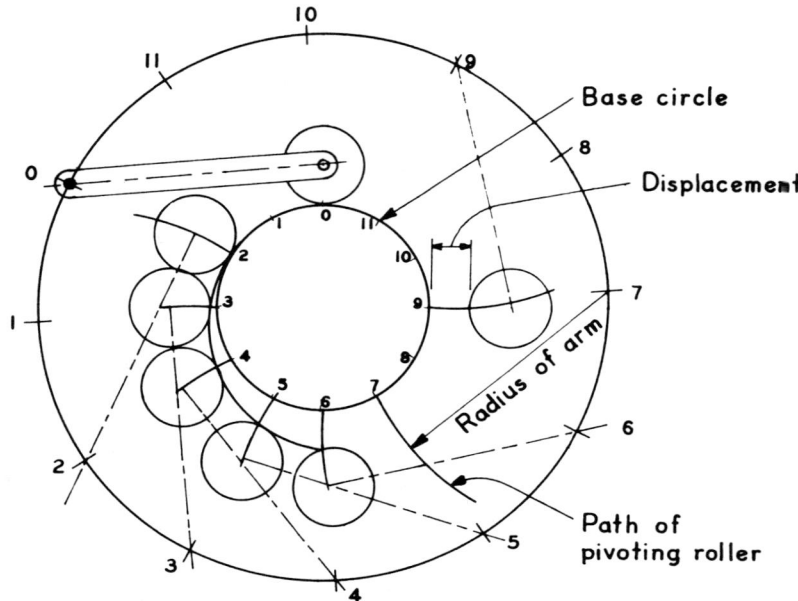

FIG. 4-20 PIVOTED FOLLOWER.

involved in a follower without offset. If the pressure angle is too large, excessively large forces will be set up in the follower and the follower may even jam against the cam contour.

The jamming problem may be understood by an analysis of the forces shown in Fig. 4-21. The normal force is resolved into components $N\cos\alpha$ and $N\sin\alpha$, where N is the normal force between cam and follower. F is the load against which the follower acts, and, R and R_2 are the reactions of the guides or bearings on the follower. Summing vertical and horizontal forces and also moments at the point of application of R_1:

$$\Sigma F_x = 0 \qquad\qquad R_1 - R_2 - N\sin\alpha = 0$$
$$\Sigma F_y = 0 \qquad\qquad N\cos\alpha - \mu(R_1 + R_2) - F = 0$$
$$\Sigma M_{R_1} = 0 \qquad \frac{d}{2}F + \mu dR_2 + aN\sin\alpha - \frac{d}{2}N\cos\alpha - bR_2 = 0$$

where μ = coefficient of friction.

To obtain an expression for the pressure angle, first eliminate R_1 and R_2 as follows. $R_1 = N\sin\alpha + R_2$. Substitute this value for R_1 in the equation for $\Sigma F_y = 0$:

$$N\cos\alpha - F - \mu(N\sin\alpha + 2R_2) = 0$$

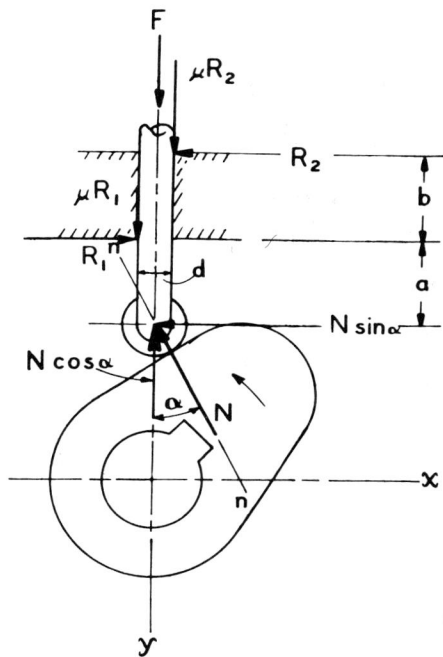

FIG. 4-21 FORCE RELATIONSHIPS FOR PRESSURE AN-
GLE.

Divide by μ:

$$\frac{N}{\mu}\cos\alpha - \frac{F}{\mu} - N\sin\alpha - 2R_2 = 0$$

Then

$$R_2 = \frac{N}{2\mu}\cos\alpha - \frac{F}{2\mu} - \frac{N}{2}\sin\alpha$$

Substitute this value of R_2 in the equation for $\Sigma M_{R_1} = 0$:

$$\frac{d}{2}F + \left(\frac{N}{2\mu}\cos\alpha - \frac{F}{2\mu} - \frac{N}{2}\sin\alpha\right)$$

$$(\mu d - b) + aN\sin\alpha - \frac{d}{2}N\cos\alpha = 0$$

When multiplied out, $\frac{dF}{2}$ cancels out and this equation becomes

$$\frac{Nd}{2\mu}\cos\alpha - \frac{N\mu d}{2}\sin\alpha - \frac{Nb}{2\mu}\cos\alpha$$

$$+ \frac{Fb}{2\mu} + \frac{Nb}{2}\sin\alpha + aN\sin\alpha - \frac{dN}{2}\cos\alpha = 0$$

Multiply through by μ:

$$\frac{Nd\mu}{2}\cos\alpha - \frac{Nd\mu^2}{2}\sin\alpha - \frac{Nb}{2}\cos\alpha$$
$$+ \frac{Fb}{2} + \frac{Nb\mu}{2}\sin\alpha + a\mu N\sin\alpha - \frac{dN\mu}{2}\cos\alpha = 0$$

Consider the possible values of $(Nd\mu^2)/2\sin\alpha$. A value of 0.3 for μ would be extremely high, therefore $\mu^2/2$ must be less than 0.5. Also d is small and $\sin\alpha$ is less than 1. Then $(Nd\mu^2)/2\sin\alpha$ is small and can be ignored.

$$N\left(\frac{\mu d}{2}\cos\alpha - \frac{b}{2}\cos\alpha\right.$$
$$\left. + \frac{b\mu}{2}\sin\alpha + \mu a\sin\alpha - \frac{d\mu}{2}\cos\alpha\right) = -\frac{Fb}{2}$$
$$N\left(-\frac{b}{2}\cos\alpha + \frac{b\mu}{2}\sin\alpha + \mu a\sin\alpha\right) = -\frac{Fb}{2}$$
$$\frac{N}{F} = \frac{b}{b\cos\alpha - \sin\alpha(\mu b + 2\mu a)}$$

If there is jamming, N will be extremely large, and for this to be so then $b\cos\alpha - \sin\alpha(\mu b + 2\mu a)$ must be extremely small. Take

$$b\cos\alpha - \sin\alpha(\mu b + 2\mu a) = 0$$
$$\frac{\sin\alpha}{\cos\alpha} = \tan\alpha = \frac{b}{\mu(2a+b)}$$

when $N = \infty$ (jamming). This expression gives the limiting value for the pressure angle α. Note that α is critically dependent on μ. Consider an adverse cam design with $a=2$, $b=1$, and $\mu=0.2$. For these values $\tan\alpha = 1.0$ and $\alpha = 45°$. The usual rule of thumb is to keep the pressure angle at or below 30°, and this would appear to be a useful limit.

Should the cam profile be too steep, any of the following adjustments may solve the problem:

1. Increase the base circle diameter. This makes the cam larger and its profile less steep. However cam mechanisms are expected to be compact.

2. Increase the angle of cam rotation required for the lift.

3. Decrease the follower rise by redesigning the mechanism.

4. Use an offset follower.

5. Change the type of follower motion.

For a given pressure angle, the performance and effectiveness of the follower will be improved by the following:

1. Low coefficient of friction in the follower guides
2. Reduced overhang, dimension a in Fig. 4-21
3. Adequate bearing length b

4.7. DRUM CAMS.

There are circumstances when it is desirable for the direction of follower motion to be parallel to the rotational axis of the cam. This is possible with a cylindrical cam or drum cam, as in Fig. 4-3. The follower is guided by either a groove in the cylinder or a raised land on its surface. Any type of follower motion, including a dwell, can be provided by a drum cam.

4.8. ECCENTRICS.

An eccentric is a circular cam rotating about an eccentric axis, as in Fig. 4-22. A flat, roller, or pivoted follower may be used. The eccentric provides simple harmonic follower motion. Important advantages of this cam are its ease of design and manufacture and its inherent accuracy of profile.

To graph the displacement of an eccentric, see Fig. 4-23. First draw a circle about the cam rotational axis with a radius equal to the eccentricity e. Divide this circle into the required number of cam stations. Draw arcs equal to the cam radius R from each station. The displacement diagram is constructed by transferring displacements horizontally as shown in the figure. Displacements may be calculated from the equations for simple harmonic motion.

Example.   The eccentric of Fig. 4-22 rotates at 120 rpm. What is the maximum value of the acceleration?

Solution.   Since the motion is harmonic,

$$a = e\omega^2$$
$$= 1.2\left(\frac{120}{60}2\pi\right)^2$$
$$= 189.5 \text{ cm/s}^2$$

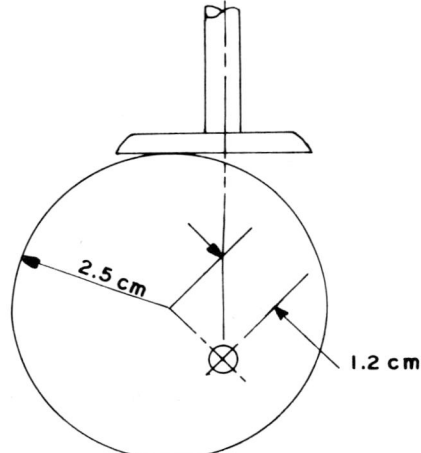

FIG. 4-22   ECCENTRIC.

Considering the relatively low speed of the cam, this is an impressive acceleration. The force required to generate such an acceleration depends on the mass accelerated and may perhaps not be large, but if exerted against a small cam area it might represent a high pressure, with the possibility of rapid wear. Note also that a doubling of the rotational speed produces four times the acceleration and also four times the force and the pressure at the cam.

Cam accelerations and forces cannot be assumed to be inconsequential.

4.9.   CONSTANT-DIAMETER CAMS.

A double-contact or constant-diameter cam, such as the one shown in Fig. 4-24, provides positive contact between cam and follower throughout the cam cycle, though a small working clearance must be allowed. The width across the cam must, of course, be constant over 360 degrees of rotation, otherwise there cannot be double contact throughout the cycle. Such cams are usually made of circular arcs, a feature that makes for ease of layout and manufacture.

The cam of Fig. 4-24 is designed as follows. Let angle δ be the desired dwell angle. Then $180° - \delta$ is the cam angle of the two motion periods of the cycle. The lift is $AQ - AP$.

A triangle ABC is constructed, with the angle at A equal to the required dwell angle δ. Sides of the triangle b and c are given

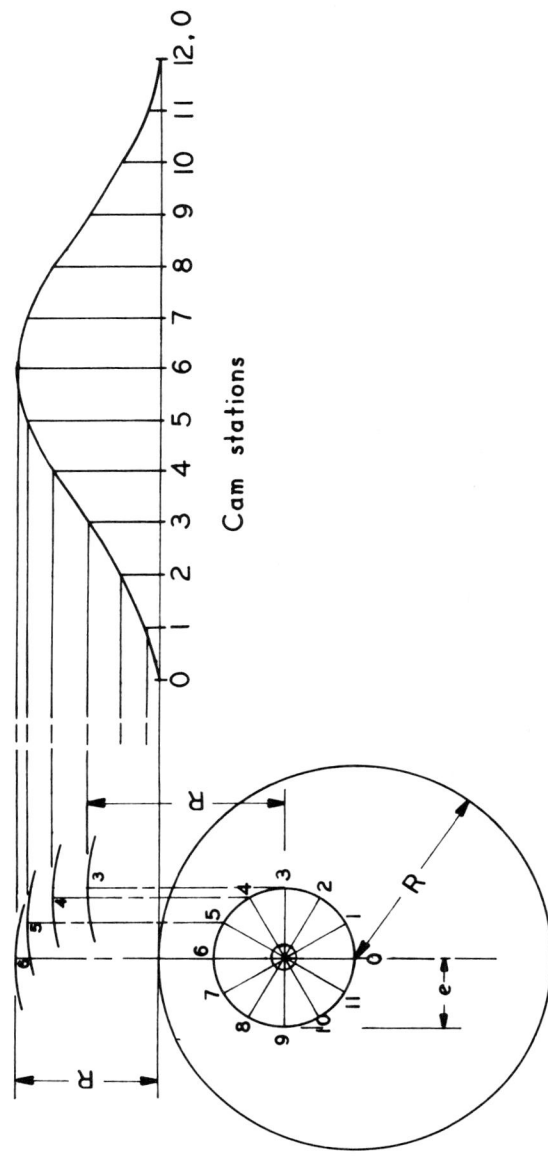

FIG. 4-23 LAYOUT OF DISPLACEMENT DIAGRAM OF AN ECCENTRIC.

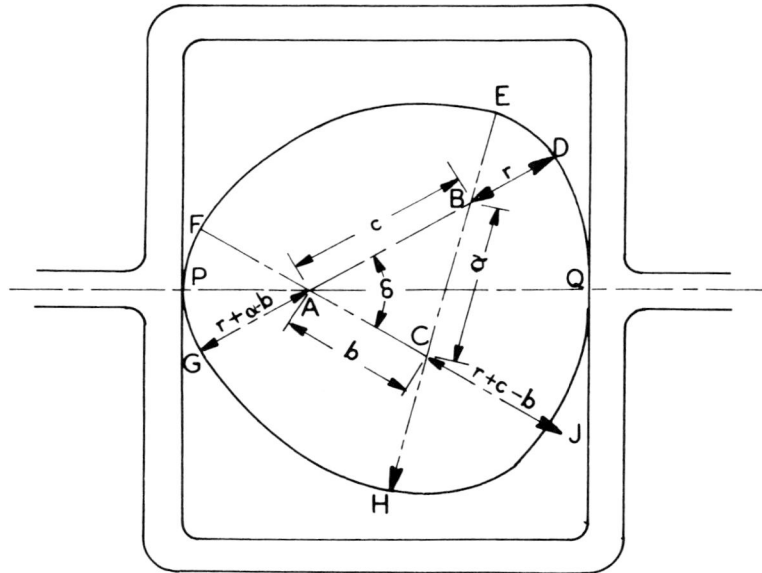

FIG. 4-24 CONSTANT-DIAMETER CAM, GENERAL METHOD OF CONSTRUCTION. THE AXIS OF ROTATION IS TAKEN AS AT A, BUT MIGHT ALSO BE LOCATED AT B OR C.

dimensions such that the length of the lift of the follower yoke equals $b + c - a$. At the vertex of the smallest angle of this triangle, which in the figure happens to be B, draw arc DE with B as center and any suitable radius r. Using C as center, draw arc EF tangent to arc ED, and with A as center draw arc FG. Arc GH has its center at B; arc HJ has its center at C. Finally, arc DJ has its center at A.

Angles B and C are made unequal in the figure. They may be made the same. If unequal, the time of advance of the follower yoke is different from the time of return. But if angles B and C are the same, the advance and return strokes are identical. It may be noted too that the rotational axis of the cam could be located at B or C, as well as at A.

If parallel tangents are drawn at two points on opposite sides of the cam, they will be separated by a distance equal to $2r + a + c - b$. This distance must, of course, be a constant distance for all parallel tangents, so that the yoke has two opposite contact points throughout the cycle.

To use an example in inches, suppose that the cam requires a dwell angle of $60°$ and a lift of 1.0 in. Then $b = c - a = 1.0$. Suppose the following lengths are selected:

$$b = \frac{15}{16} \text{ in.,} \qquad c = \frac{21}{16} \text{ in.}$$

Then
$$a = \frac{20}{16} \text{ in.}$$

Next suppose r is taken as $10/16$ in. Then the major axis of the cam has a length of

$$r + a - b + c + r = \frac{10+20-15+21+10}{16} = \frac{46}{16} \text{ in.,}$$

which is the constant diameter. From the figure the lift

$$= (c+r) - (r+a-b)$$
$$= \frac{(21+10)-(10+20-15)}{16}$$
$$= 1.0 \text{ in.}$$

A constant-diameter cam of the type used in sewing machines and motion picture mechanisms is shown in Fig. 4-25. This cam uses two radii r_1 and r_2 as shown, with a dwell angle δ. The active portions of the cam profile are constructed with arcs drawn from A and B with a radius of $r_1 + r_2$. The corners at A and B are suitably modified for a smoother action.

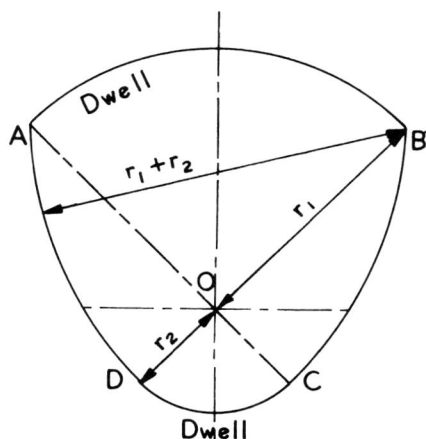

FIG. 4-25   CONSTANT-DIAMETER CAM CONSTRUCTED FROM TWO RADII.

4.10. ACCELERATION OF A FOLLOWER.

The following example serves as an illustration of the method of determining the acceleration in a cam follower. The cam of Fig. 4-26 rotates clockwise at 180 rpm. The total acceleration of the follower is to be determined at its top dead center position, which is the position shown. At this position the follower has an instantaneous tangential velocity of zero.

The acceleration of Q can be determined from the relationship

$$a_Q = a_P \mapsto a_{Q/P}$$
$$a_{Q_n} \mapsto a_{Q_t} = a_{P_n} \mapsto a_{P_t} \mapsto a_{Q/P_n} \mapsto a_{Q/P_t}$$

But $\qquad a_{Q_n} = 0$ since $v_Q = 0$

and $\qquad a_{P_t} = 0$

Then $\qquad a_{Q_t} = a_{P_n} \mapsto a_{Q/P_n} \mapsto a_{Q/P_t}$

$$a_{P_n} = 2.5 \left(\frac{180}{60} 2\pi \right)^2$$
$$= 887 \text{ cm/s}^2$$

a_{Q/P_n} can be determined from $\dfrac{(v_{Q/P})^2}{2.75}$, and $v_{Q/P}$ can be determined

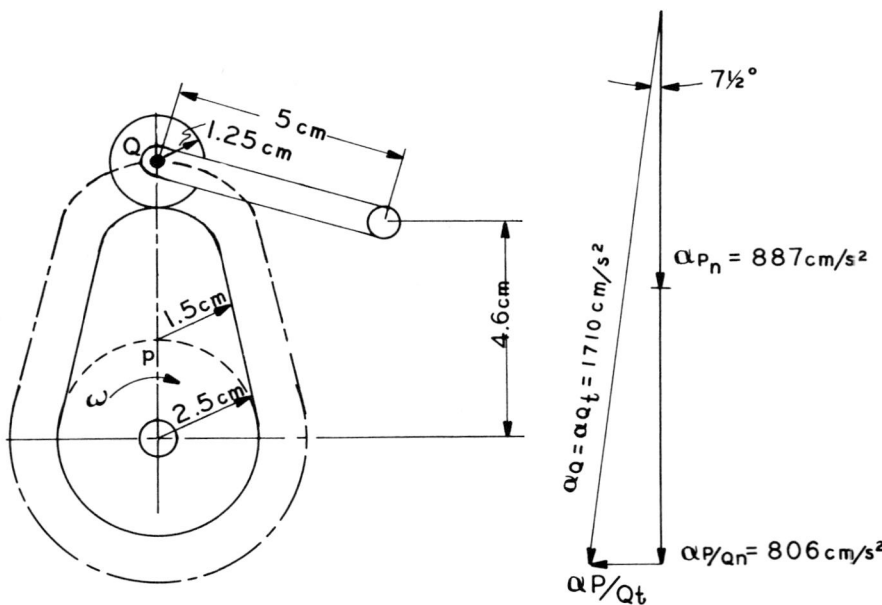

FIG. 4-26 ACCELERATION OF A FOLLOWER.

from the vector relationship

$$v_Q = v_P \mapsto v_{Q/P}$$

But $\quad v_Q = 0$

$$v_P = 2.5\frac{180}{60}2\pi = 47.1 \text{ cm/s}$$

Then $\quad v_{Q/P} = 47.1 \text{ cm/s, and } a_{Q/P_n} = \frac{(47.1)^2}{2.75} = 806 \text{ cm/s}^2$

Of the four acceleration vectors, two are completely determined in magnitude and direction. For the remaining two acceleration vectors only the direction is known. The information at hand is sufficient to solve completely the acceleration vector polygon, which is drawn in Fig. 4-26. Despite the slow speed of the cam, the acceleration of the follower is high, 1710 cm/s².

4.11. SYNCHRONIZING OF CAMS.

The two cams of Fig. 4-27 drive ON-OFF switches in an electrical contact system. Each cam has two long dwells, one for ON (follower on base circle), and one for OFF (follower at maximum lift). When one cam gives the OFF position the other cam must give

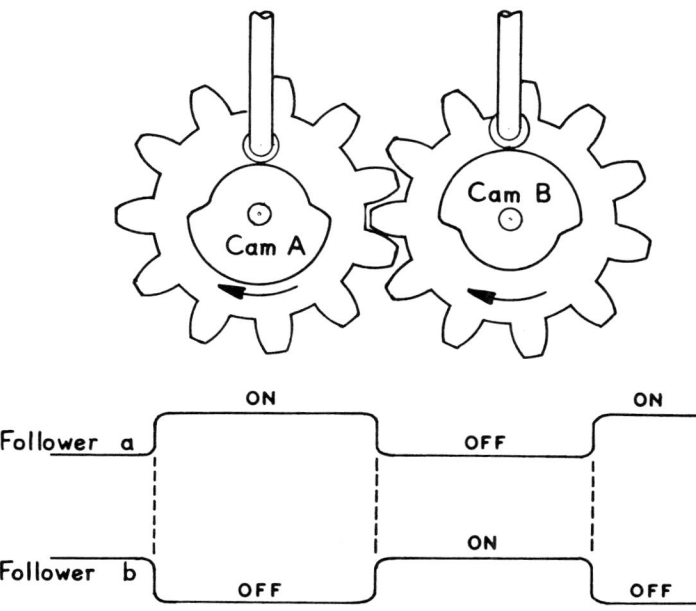

FIG. 4-27 SYNCHRONIZED ON-OFF CAMS.

the ON position. The two dwell angles on each cam must be identical and approximately 180°. Since the cams must be synchronized, they are coupled by gears.

If an overlapping cam action is required, where one cam must switch to ON before the other cam switches to OFF, it is necessary only to rotate one cam through the required number of degrees of overlap so that the two cams are out of phase by less than 180°. An alternate method is to alter the dwell angle of one cam.

If the diameter of the gear connected to cam *A* in Fig. 4-27 is made twice that of the gear connected to cam *B*, then cam *B* will rotate twice as fast as cam *A*. Cam *B* would then give a complete ON-OFF cycle in the time that cam *A* made an ON or an OFF cycle. But the same double cycle in cam *B* could be produced by a 1:1 speed ratio if cam *B* is profiled for two ON dwells and two OFF dwells.

The follower relationships of Fig. 4-27—one cam OFF while the other cam is ON—can be produced by one cam with two opposed followers, as in Fig. 4-28. When using two independent followers on the same cam, overlap is produced by placing one follower at an angle other than 180° to the other follower. If a large overlap is required, such that the followers will interfere with each other, the cam must be made wide enough for the placing of the two followers in side-by-side position.

When synchronizing cams by means of gears, it is possible to forget that the two gears in a pair rotate in opposite directions.

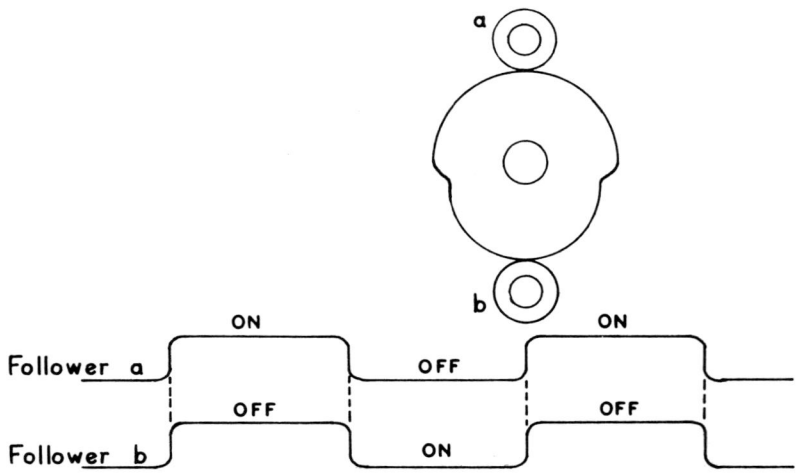

FIG. 4-28 CAM WITH TWO INDEPENDENT FOLLOWERS.

4.12.   THE GENEVA MECHANISM.

This mechanism is a well-known intermittent motion and index-ing device. The Geneva mechanism of Fig. 4-29 shows six slots in the Geneva wheel W; actually any number of slots may be used, with a minimum of three. The number of slots determines the speed ratio between driving wheel D and Geneva wheel W. With six slots, W makes $\frac{1}{6}$ revolution when D makes 1 revolution.

The driver D revolves at constant rotational speed and carries a pin turning with it that engages with the slots in W. At the instant of engagement of the pin with the slot, the radial axis of the pin must be at right angles to the axis of the slot in order to avoid impact at engagement. Therefore, at the instant of engagement, and of disengagement, the pin does not drive the Geneva wheel. As the pin enters the slot the angular velocity of the Geneva wheel in-creases, and as the pin completes its passage down the slot the angular velocity of W falls to zero. The driving wheel is cut away, as shown, so as to clear the slotted lobe of W while W turns. At the instant the pin leaves the slot, the circular arc of D engages the matching curved surface of W between the slotted lobes, thus lock-ing W until the pin engages the next slot.

Fig. 4-30 shows the dimensional relationships of a Geneva mechanism at the instant of engagement. The rotation angle of the Geneva wheel $= \theta = 360°/n$ where $n =$ number of slots.

$$a = L_1 \cos \alpha,$$

where $L_1 =$ length of the crank arm to the center of the pin.

$$b = L_o - L_1 \cos \alpha$$

The maximum value of α is $180°/n$, therefore

$$L_1/L_o = \cos(180°/n)$$

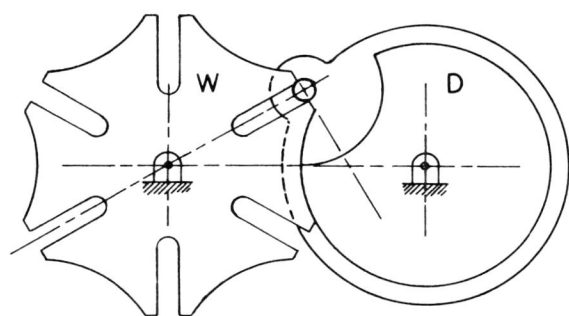

FIG. 4-29   SIX-POSITION GENEVA MECHANISM.

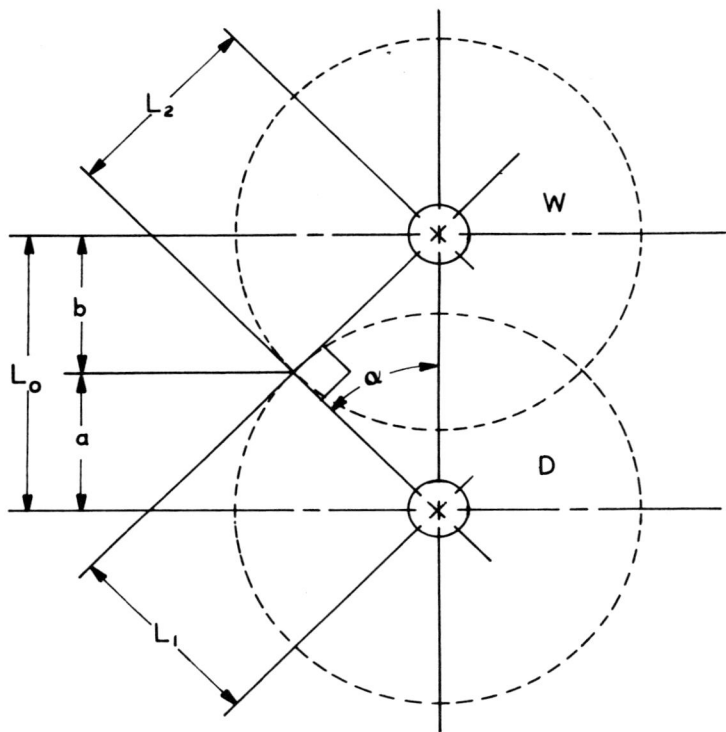

FIG. 4-30   DIMENSIONAL RELATIONSHIPS IN A GENEVA
MECHANISM.

Therefore L_o, the distance between the axes of the two wheels

$$= \frac{L_1}{\cos \dfrac{180°}{n}}$$

To make a graph of the angular velocity of a Geneva wheel for
one indexing cycle, a standard method for mating wheels is used. See
Fig. 4-31. In the figure, the line of centers is the line WD, where W
is the axis of rotation of the Geneva wheel and D is the axis of
rotation of the driver. The circular path of the pin between engagement with the slot at point 0 to disengagement at point 10 is shown,
this curve being divided into 10 angular increments. At the instant of
engagement and disengagement the line D to 0 (axis of the crank
and pin) must be perpendicular to the line W to 0 (axis of the slot).
The lines W to 1, to 2, to 3, etc. are successive positions of the axis of
the slot.

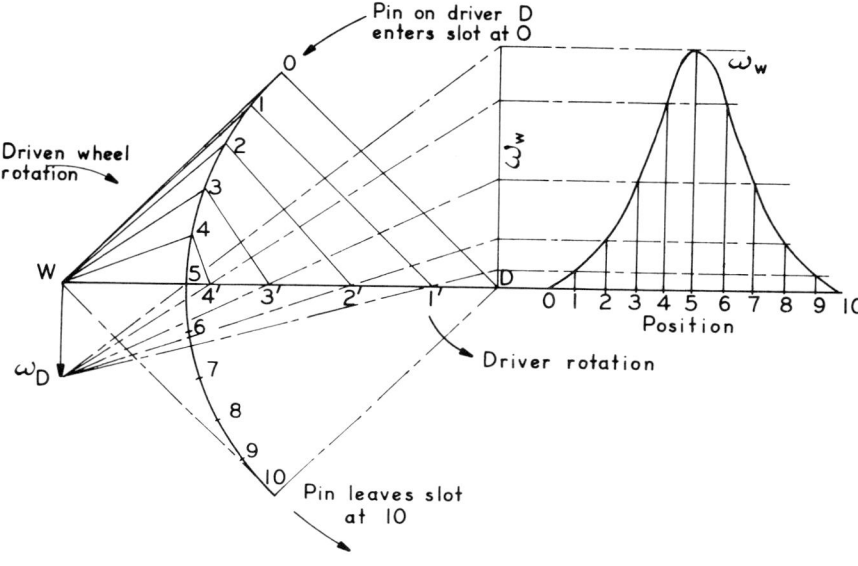

FIG. 4-31 VELOCITY ANALYSIS OF A GENEVA MECHANISM.

To determine values of the angular velocity of the Geneva wheel, we use a principle that is perhaps not intuitively very obvious, and which will be discussed further in the following chapter: the angular velocities of a driver and a follower are inversely proportional to the segments into which the common normal cuts the line of centers. The line of centers is, of course, the line WD. Since the pin is sliding along the slot, the common tangent of the two wheels is the axis of the slot, and the common normal at any point along the slot must be at right angles to the slot. A normal is by definition a line at right angles to a tangent. Lines $W0$, $W1$, $W2$, $W3$, etc. therefore are common tangents. The lines $11'$, $22'$, $33'$, etc. are drawn at right angles to the slot at the several positions and are normals.

The principle states that angular velocities are inversely proportional to the segments into which the line of centers is divided by the normal. To take an example, consider the normal $22'$, intersecting the line WD at $2'$. It divides WD into two segments, $W2'$ and $2'D$. The angular velocity of the Geneva wheel is proportional to the length $2'D$ if the speed of the driver is proportional to the length $W2'$. Suppose $W2'$ is 40 units long, and $2'D$ is 20 units long. Then the angular velocity of the Geneva wheel is $\frac{20}{40}$ of the angular velocity of the driver.

Since the segment of the line of centers at the W end of the line represents the speed of the driver, because of the inverse relationship, the angular velocity of the driver, ω_D, is laid out vertically from W to some convenient scale. Lines are run from the tip of this angular velocity line through 5′, 4′, 3′, 2′, etc. to intersect a line drawn vertically from D, as shown. These intersections on the line drawn up from D are the instantaneous angular velocities of the Geneva wheel to the same scale as used for ω_D. This construction uses the principle of similar triangles to convert the segments of the line of centers to a scale based on ω_D.

Another approach to understanding the motion of a Geneva mechanism is to convert it into the following kinematically equivalent linkage. The term "kinematically equivalent" means that the substitute linkage has exactly the same motion as the Geneva mechanism. This substituted linkage applies for any number of slots of the Geneva wheel.

In Fig. 4-32 a Geneva mechanism is shown with the pin P at some point along the slot. Point W is the axis of rotation of the Geneva indexing wheel; D is the axis of rotation of the driving wheel. The Geneva mechanism is converted into the kinematically equivalent linkage imposed on the mechanism: a driven arm PW, a driving arm DQ, and a connecting link PQ. For kinematic equivalence, link PQ must be at right angles to PW and driving arm DQ must be at right angles to PQ. The driving and driven arms must rotate about the axes of rotation of the wheels. The speed ratio for the instantaneous position shown is given by the ratio of the linkage lengths PW/DQ, and the instantaneous angular velocity of the driven wheel is PW/DQ times the angular velocity of the driver. This construction can be applied at any position of the pin in the slot for the purpose of determining the instantaneous velocity—three links all at right angles, with the end links rotating about the axes of

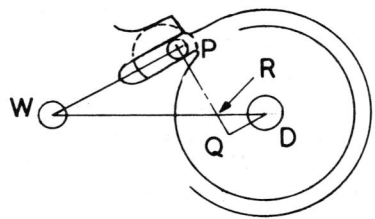

FIG. 4-32 KINEMATIC EQUIVALENCE FOR A GENEVA MECHANISM.

rotation of the two wheels. As different positions are adopted, the lengths of the three imaginary links will change, as they must, because the speed ratio of the two wheels is not constant. When the pin lies on the line of centers, the length of the connecting link becomes zero.

Note that the imaginary connecting link is the common normal. The speed ratio can be found from the lengths of the imaginary links, or from the lengths RD and WR into which the line of centers is divided, because triangles DQR and WPR are similar triangles.

PROBLEMS

1. Design flat cams to the following specifications. Check that pressure angles do not exceed 30°. Draw displacement diagrams. For any flat-face followers, determine a suitable width or diameter for the follower.

   a) Base circle 6.25 cm diameter, roller follower 1.25 cm diameter, not offset, clockwise rotation of the cam

Cam Angle	Lift	Motion
0–180°	0–2.5 cm	constant acceleration
180°–210°	dwell	–
210°–330°	2.5 cm to 0	constant deceleration
330°–360°	dwell	–

   b) Base circle diameter 6 cm, counterclockwise rotation of cam with flat-face follower (not offset)

Cam Angle	Lift	Motion
0–90°	0–2.5 cm	cycloidal
90°–180°	dwell	–
180°–330°	2.5 cm to 0	harmonic
330°–360°	dwell	–

   c) Base circle diameter 5 cm, counterclockwise rotation, roller follower 1.25 cm diameter with 1 cm offset

Cam Angle	Lift	Motion
0–180°	0–2 cm	harmonic
180–360°	2 cm–0	modified trapezoidal

2. Design a flat cam with radial roller follower to operate at 2000 rpm. Base circle diameter 5 cm, lift 1.25 cm. Lift and fall are to be accomplished in 120 degrees each, separated by 60° dwells. Since this is a high-speed cam, cam motion must have controlled acceleration. Rotation is to be clockwise, and follower roller diameter is to be selected by the designer.

3. Determine the maximum acceleration for the cam follower of Problem 2.

4. Design a drum cam to provide the follower motion of Problem 1 (a). The groove in the drum is to be cut from the drawing of the displacement diagram, using a drum diameter of 10 cm. Supply any other necessary dimensions.

5. Determine the maximum velocity of the followers for Problems 1(a) and 2. The cam of Question 1(a) rotates at 120 rpm.

6. Design a slow-speed cam with radial roller follower for clock-wise rotation, to provide the following motion. Use a base circle of 5 cm diameter and a roller 1.5 cm in diameter.

Cam Angle	Lift	Motion
0–90°	0–2.5 cm	uniform velocity
90°–180°	2.5–1.5 cm	uniform velocity
180°–270°	1.5–2.5 cm	uniform velocity
270°–360°	2.5 cm–0	uniform velocity

7. Completely design a high-speed cam with pivoted roller follower to move the follower over a total swing of 30° (the linear displacement of the follower is not significant). The only requirements are two:
   a) When the follower is on the base circle its arm must be horizontal. This is the initial position of the cam cycle.
   b) The cam components must make as compact a mechanism as possible.

8. Determine the total acceleration of the follower of Problem 7 when its arm is horizontal.

9. A translating roller follower is to have a lift of 2.5 cm. At its lowest position, the center of the roller must be located 6.25 cm above the axis of rotation of the cam. The roller diameter is to be 2.0 cm. Design an eccentric cam to produce the follower motion.

10. The figure shows an eccentric cam with a pivoted follower. Determine the highest and the lowest position of the follower roller. The diameter of the roller is 2.0 cm.

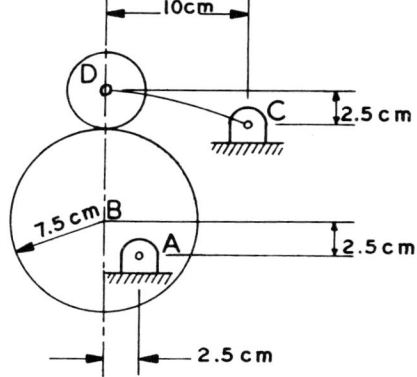

FIG. P4-10   ECCENTRIC WITH PIVOTED FOLLOWER.

11. The eccentric in the figure has a diameter of 10 cm and rotates at 1 rps counterclockwise. Determine:
   a)   The velocity of point *P* at the position shown
   b)   The velocity of sliding between eccentric and follower

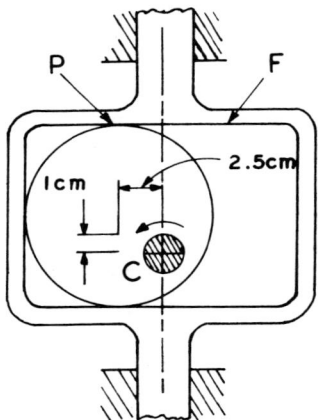

FIG. P4-11   ECCENTRIC WITH YOKE.

12. For the eccentric cam of the figure, graph the displacement and determine the maximum pressure angle.

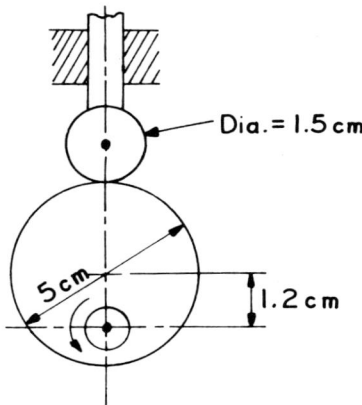

FIG. P4-12   ECCENTRIC AND RADIAL ROLLER.

13. Design a cam to operate a microswitch. A movement of 1.5 mm minimum is required to activate the switch. The cam must rotate at 30 rpm. The cycle is:
   OFF—1.25 sec (follower on base circle)
   ON—0.75 sec

Use as rapid a rise and fall as possible, subtracting the rise from the allowed 1.25 sec OFF time and the fall from the allowed 0.75 sec ON time.

14. Provide the following cycle for the cam layout of the figure.

Cam Angle	Lift at Point P	Motion
0–120°	0–2.5 cm	uniform acceleration
120°–150°	dwell	–
150°–210°	2.5–1.5 cm	uniform acceleration
210°–240°	dwell	–
240°–300°	1.5 cm–0	uniform acceleration
300°–360°	dwell	–

The follower lift may be taken along the chord of the arc of motion. Determine also the face length of the follower.

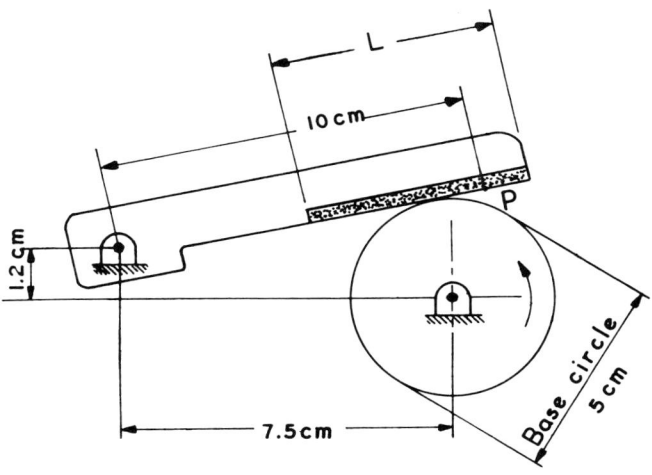

FIG. P4-14   PIVOTED FOLLOWER.

15. A cam is to provide the following motion to an offset roller follower:

Time	Lift	Motion
0–0.5 sec	2 cm	harmonic
0.5–0.75 sec	dwell	–
0.75–1.0 sec	2 cm–0	harmonic
1.0–1.25 sec	dwell	–

The final dwell time is not critical, but must lie within the range of 0.1 to 0.25 second. Designer to select the cam rpm.

Use an offset of approximately one quarter of base circle diameter and a pressure angle not in excess of 30° maximum. Design the cam completely, including rpm and direction of rotation.

16. Design a constant-diameter cam of the type shown in Fig. 4-25 to provide a follower lift of 1.5 cm. Draw the displacement diagram.

17. Design a constant-diameter cam of the type shown in Fig. 4-24 to provide a follower lift of 3 cm. Draw the displacement diagram.

18. Two synchronized cams are to operate ON-OFF cycles for electric switches. Switch action requires a minimum movement of 0.8 cm of upward movement. The two cams are to be geared together. The two switches operated by the two cams are to be switched ON simultaneously. The cycle for one cam is to hold the ON position for 1 second, then hold OFF for 1 second. In this 2-second cycle the other cam must switch OFF twice, using equal ON and OFF times. Design the cam system.

19. A single cam has two followers, both of which operate limit switches. While one limit switch is ON, the other limit switch is OFF. The ON part of the cycle of one switch must overlap the OFF part of the cycle of the other switch so that at all times one switch is ON. Design the cam system, assuming low-speed operation. Limiting switch operation to ON requires vertical movement of 0.8 cm minimum.

20. For the Geneva motion shown in the figure, graph the angular velocity, and from this graph draw the graph of angular acceleration for one indexing cycle. Driver rotates at 1 rad/s.

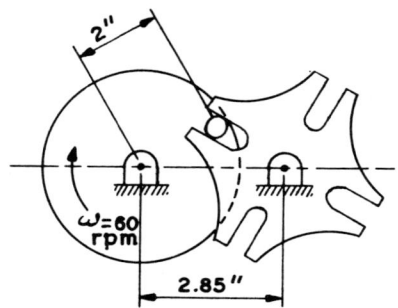

FIG. P4-20   GENEVA MECHANISM.

21. Design a Geneva mechanism for low output torque using 5 slots and a center distance between shafts of 15 cm.

chapter five

Gearing

5.1. SPEED RATIO.

A number of power transmission devices serve the purpose of transmitting rotary motion from one shaft to another. The range of such devices includes mating gears, power chain and sprockets, V-belts and pulleys, and others. If considerable torque must be transmitted, and also if there must be a fixed and unvarying ratio of angular velocities between the two shafts, then gearing must be selected for the purpose. If the speed ratio between the two shafts must be, for example, 1:3, this ratio must be unvarying for any amount of rotation. That is, if the driver shaft rotates 10 revolutions, then the driven shaft must rotate 30 revolutions; if the driver rotates 3°, then the driven shaft must rotate 9°, and so for any rotational angle. Though it would appear that a belt or a chain drive can provide such fixed speed ratios, they cannot. A belt can slip and creep, and a power chain has a speed variation called chordal action when a link of the chain engages the sprocket tooth. If such small speed variations are acceptable, as they are for many drives, then belts or chain can be employed. Gears do not have such speed ratio deficiencies.

A pair of cylinders in pure rolling contact will provide an unvarying speed ratio, provided that the transmitted torque is not excessive. With no slip, then the speed ratio is governed by the diameters of the cylinders:

$$\frac{\omega_1}{\omega_2} = \frac{D_2}{D_1}$$

If gears are capable of exact and unvarying speed ratios, then their tooth form must be suited to this exacting requirement.

5.2. CONJUGATE CURVES.

If two friction cylinders are replaced by two mating gears to provide the same speed ratio, then the curve of the tooth profile is subject to some restrictions. There are a number of suitable curves for this purpose, though the involute is almost always used.

When the contacting surfaces of two rotating bodies are shaped so that one drives the other with a fixed speed ratio, the two surfaces are said to be *conjugate* to one another. The profiles of such surfaces are called conjugate curves. For a fundamental understanding of conjugate action, consider Fig. 5-1.

This figure shows a driving member D turning about a shaft at O. To generalize the shape of D, its profile is made a straight line. While this is an improbable shape for conjugate action, such action can be produced from a straight line. See also the flat gear of Fig. 5-5, which also produces conjugate action. The driven member in Fig. 5-1, called G, is rotating about a shaft at Q. If G is driven by the flat surface of D at a constant angular velocity, the profile of its surface

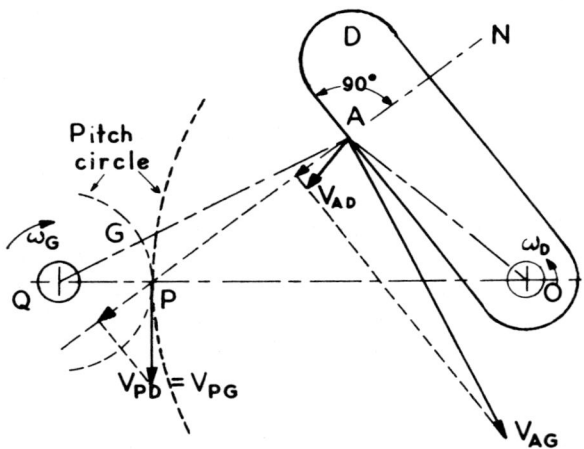

FIG. 5-1 CONJUGATE ACTION.

is initially quite unknown, and therefore cannot be shown in the figure.

The contacting surfaces of D and G must give a constant ratio of angular velocity as the two members rotate. At the instant shown in the figure, suppose that the contact point between D and G is A. The velocity of A on D is $(OA)\omega_D$, and this velocity is perpendicular to OA. Similarly the velocity of A on G is perpendicular to AQ. These two velocities have different directions, therefore sliding contact occurs at A. But as explained in Sec. 3.9 sliding is a relative velocity tangential to the surfaces at the contact point, and the components of v_{AD} and v_{AG} in the normal direction (direction NN) are equal. From this construction, the magnitude of v_{AG} is determined.

But D and G must have a constant speed ratio. They must engage, therefore, as if both were friction cylinders. As in the case of a gear or a sprocket, these imaginary cylinders are called *pitch circles*; their diameters are called *pitch diameters*, and the point of tangency of these pitch circles is called the *pitch point*. This pitch point lies at the intersection of NN, the common normal to D and G at the contact point A, and the line of centers, which is QO. This intersection is designated P in the figure. Body D must be imagined as expanded to include its pitch circle. The velocity of P on D must be perpendicular to QO and must have the same component in the direction NN as point A if D is a rigid body. Similarly, P on the pitch circle of body G has a velocity that is perpendicular to QO and the same component along NN as point A. That is, $v_{PG} = v_{PD}$ because they have the same component and the same direction. The two pitch circles roll without slip because of identical contact velocities. Point P is the only point on QO where these velocities v_{PG} and v_{PD} are identical.

The pitch point P determines the ratio of angular velocities.

$$\frac{\omega_G}{\omega_B} = \frac{OP}{QP}$$

This ratio can be preserved only if the pitch point remains at P throughout one revolution of either mating body. In Fig. 5-2 body D turns to provide a different point of contact between D and G, point B. This new point B must lie on a normal to D through P as did A. The common normal to two conjugate bodies must always pass through the pitch point if the speed ratio is to be constant. This is the basis of the design of conjugate curves, and the design of a conjugate profile can now be considered.

One of two conjugate profiles is shown in Fig. 5-3 as the eccentric disk B rotating about axis O. The other profile, which must

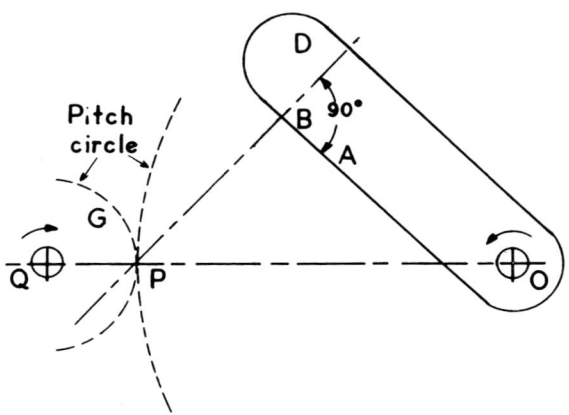

FIG. 5-2 DRIVER *D* ROTATED TO ANOTHER POSITION.

be designed, is designated *G* and rotates about an axis at *Q*. From the preceding discussion, it is known that if the two profiles are to be conjugate, then the pitch point must lie on the line *QO*. The profile *B* turns at a constant angular velocity counterclockwise. Then the profile *G* will turn clockwise. For simplicity a 1:1 speed ratio will be taken. With this speed ratio, the two profiles will have the same motion as two cylinders of equal diameter rolling on each other without slip. These idealized cylinders are shown in the figure as two pitch circles tangent at the pitch point *P*. Because of equal speed ratios, *P* must lie equidistant from *Q* and *O*.

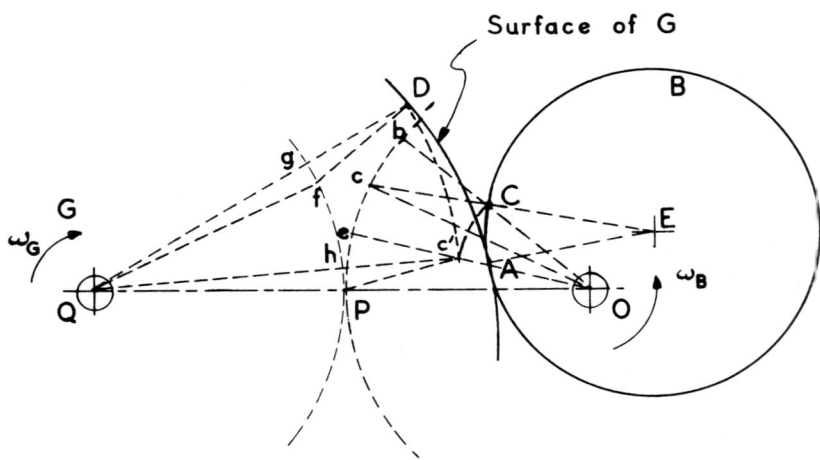

FIG. 5-3 DETERMINING A POINT ON A CONJUGATE SURFACE.

The conjugate surface of G will be designed by finding a series of points on its profile and then running a smooth curve through these points. The normals to these points must pass through P for conjugate action.

In the position of B shown in the figure, its surface contacts G at some point. This point is found from the principle that the common normal to surfaces B and G at this point of contact must pass through P. So a line is set up from P normal to surface B. A radial line is normal to a circle, so PE is normal to B, and point A is the contact point on both B and G. Thus one point on G is determined.

A second point on B, point C, is selected. The normal at C is EC. When point C turns, so that EC passes through P, C will be in contact with G. Since it is laborious to redraw B in the rotated position, the normal in Fig. 5-3 is extended to the pitch circle of B at C.

Consider the segment cCO in Fig. 5-3. Rotate this segment so that point c travels along the pitch circle to P. Side Oc becomes side OP. Point C moves along a circular arc about O to position c'. Segment cCO is now rotated to position $Pc'O$. To locate c', extend OC to the pitch circle at b, and note that arc Pe equals arc cb in length; c' is on the line eO. The normal cC is now Pc'. Therefore c' is a point of contact between G and B.

Meanwhile G has also rotated—through the same angle as B—to bring a point to the contact point c'. To find the position of this point when the two conjugate bodies were in original contact at A, remember that this second point of G has travelled a circular arc about Q to c'. To retrace the path of c' on G backward, consider the segment QPc'. The point P on this segment moves back along the pitch circle of G to f. The speed ratios of B and G are identical, therefore, this backward rotation moves through the same arc length as the corresponding point on B, that is, through the same angle. Therefore arc Pf equals arc Pc. Arc distance Ph equals arc distance fg, and point D on G is located. This point D is the original position of point c' on G. We now have two conjugate points located on the two conjugate surfaces.

This explanation may not be entirely clear when first considered. Therefore a brief explanation of the location of a third point on G is provided. This point is point L (Fig. 5-4).

A point k is selected on B. A normal (from E) is drawn through k to the pitch circle of B (point m), also a line is drawn from O to the same point m. This construction provided the triangular segment Omk, the device used in the previous explanation. This segment is rotated down to the position OPk', the point k rotating about O to k'. The triangular segment QPk' on G is then constructed. The segment

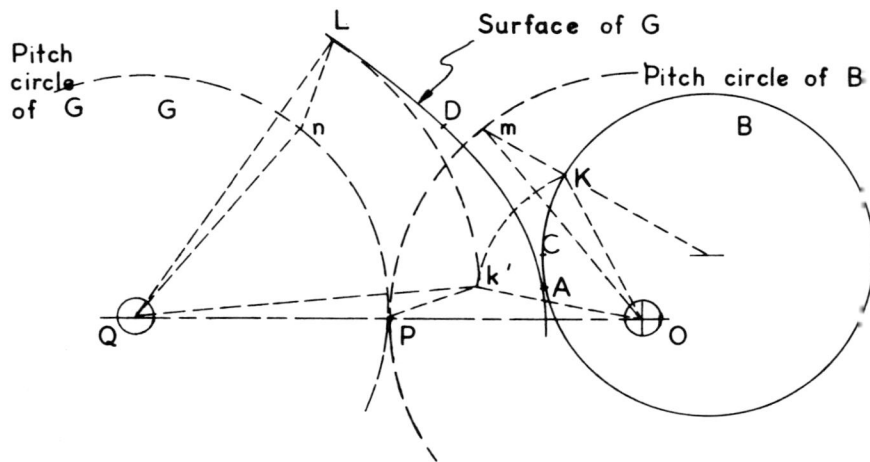

FIG. 5-4   DISK B ROTATED TO ANOTHER POSITION.

QPk' is then rotated backwards to its original position through the same arc on its pitch circle as the triangular segment *Omk* was rotated, that is, arc *mP* must equal arc *Pn*. This makes *k'* on *G* rotate back to position *L* (on *G*) about *Q*.

Another example is given in Fig. 5-5. The driving gear is a square gear with rounded corners, but is given its pitch line profile without showing the gear teeth. The two pitch circles for the mating gears are shown. A 1:1 speed ratio is selected for the sake of

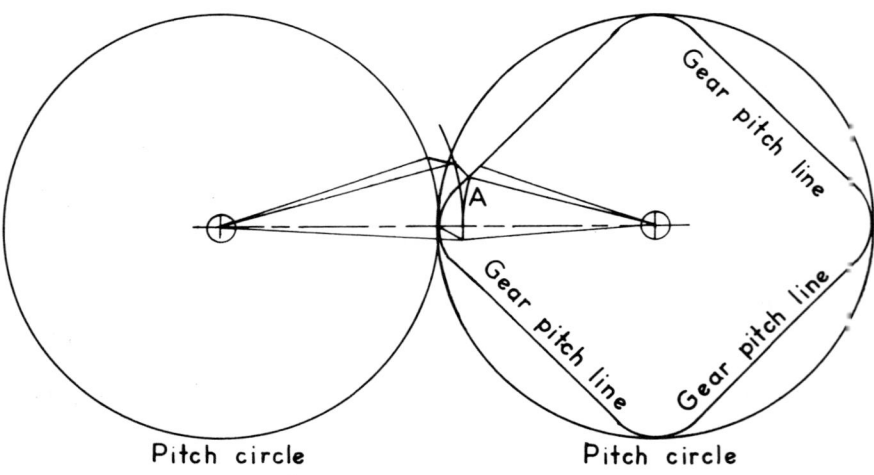

FIG. 5-5   A SQUARE GEAR.

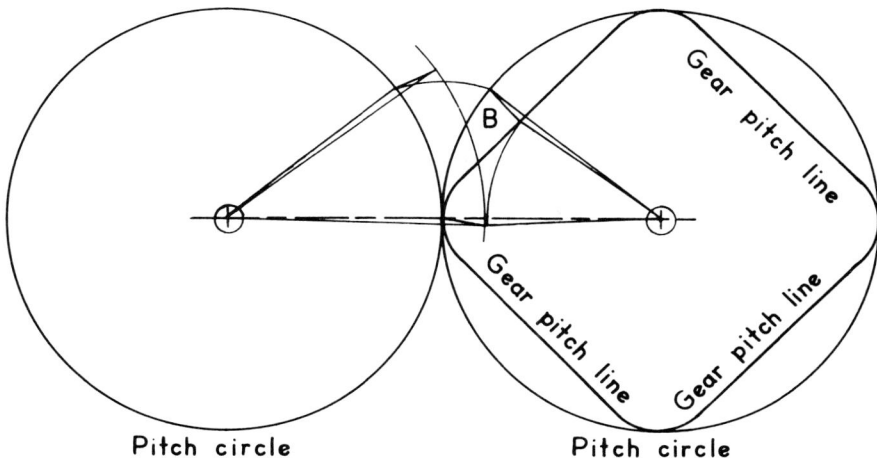

FIG. 5-6   CONJUGATE ACTION.

simplicity. The shape of the driven gear is unknown. A square gear may seem improbable; nevertheless such a device has been put to practical use.

A point *A* on the square gear is selected. The usual triangle reaching to the pitch circle of the square gear is drawn, using a normal to the flat side of the gear. The triangle is rotated so that the normal passes through the pitch point. In this position, point *A* has rotated slightly below the pitch point. The corresponding triangle on the driven gear is drawn, and is rotated back through the same length of arc on its pitch circle to locate one point on the driven gear.

The construction is repeated again for another point *B* in Fig. 5-6. The profile of the driven gear can be drawn when a sufficient number of points on it are determined.

5.3. NONCIRCULAR GEARS.

Noncircular gears (Fig. 5-7) are not used to provide a uniform angular velocity to the driven gear. Instead they are used as an alternate mechanism to cams and linkages in generating a varying angular velocity of some desired pattern in the driven gear. Such gears have been employed in computing devices, printing presses, and automatic machinery. Until the introduction of the numerically-controlled gear shaper, such gears were expensive to manufacture.

The more common noncircular gear shapes have been square, spiral, and a number of elliptical configurations. Here we will briefly

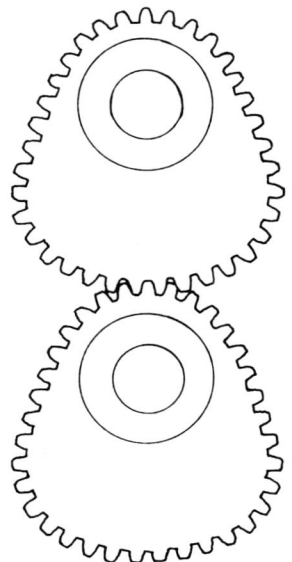

FIG. 5-7 NON-CIRCULAR GEARS.

discuss only the simplest application of elliptical gears, which is represented by Fig. 5-8.

The standard equation of the ellipse is

$$\frac{x^2}{a^2} + \frac{y^2}{b^2} = 1$$

with the origin of coordinates at the center of the ellipse. The major semi-axis of the ellipse is of length a; the minor semi-axis has a length b.

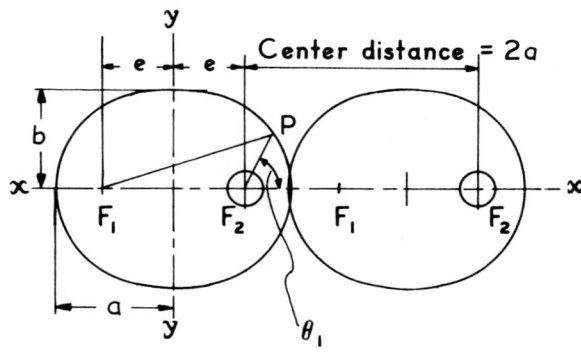

FIG. 5-8 ELLIPTICAL GEARS.

In accord with this equation, an ellipse could be drawn by fastening the ends of a cord of length $2a$ at the foci (focal points) F_1 and F_2 and passing a pencil point along the string as it is held tightly by the pencil. Then for any point P on the ellipse, the sum of $F_1P + F_2P$ is a constant and is equal to $2a$. The foci are located on the X-axis at a distance $\sqrt{a^2 - b^2}$ from the Y-axis. This distance is designated e in the figure. The *numerical eccentricity* of the ellipse ϵ is defined as

$$\epsilon = \frac{\sqrt{a^2 - b^2}}{a} \quad \text{and} \quad \epsilon = \frac{e}{a}$$

Mating elliptical gears are always twins, and the axis of rotation for an elliptical gear is always placed at a focal point. The center distance for a pair of mating elliptical gears therefore is $2a$; this length is the first design decision that must be made. The driving elliptical gear rotates at uniform angular velocity ω_1 and the angular velocity of the driven gear is given by the equation

$$\omega_2 = \omega_1 \frac{1 - \epsilon^2}{1 + \epsilon^2 + 2\epsilon \cos \theta_1}$$

where θ_1, the angle of rotation of the driven gear is the angle of the line F_2P from the major axis of the ellipse, as shown in Fig. 5-8. At $\theta_1 = 0$, the position of least angular velocity, $\cos \theta_1 = 1$ and

$$\omega_2 = \omega_1 \frac{1 - \epsilon^2}{1 + \epsilon^2 + 2\epsilon}$$

Substituting $\dfrac{e}{a} = \epsilon$, this reduces to $\omega_2 = \omega_1 \left(\dfrac{a - e}{a + e} \right)$ at $\theta_1 = 0$.
At $\theta_1 = 180°$, $\cos \theta_1 = -1$ and

$$\omega_2 = \omega_1 \frac{1 - \epsilon^2}{1 + \epsilon^2 - 2\epsilon}$$

Again by substituting $e/a = \epsilon$ this reduces to $\omega_2 = \omega_1 \left(\dfrac{a + e}{a - e} \right)$ at $\theta_1 = 180°$ the position of maximum angular velocity of the driven gear. Thus, the maximum and minimum velocities of the driven gear are reciprocals of each other.

5.4. TERMINOLOGY OF CIRCULAR GEARS.

A pair of mating circular gears will provide the same constant speed ratio as a pair of rolling cylinders. To evolve a pair of gears from a pair of rolling cylinders, the cylinders must be notched, and projections must be added between these notches to produce inter-meshing teeth. The diameters of the original cylinders are termed the *pitch diameters* of the gears. When the two gears mate, then the two pitch circles are tangent to each other at a point called the *pitch point*. The pitch point lies on the line through the centers of the two gears, the *line of centers*. The speed ratio, as for a pair of mating cylinders, is govered by the pitch diameters:

$$\frac{\omega_1}{\omega_2} = \frac{D_2}{D_1}$$

Other terminology for gears is shown in Fig. 5-9.

Circular Pitch, p_c. The distance between the same points on adjacent teeth, measured on the pitch circle. The sum of tooth width plus space width on the pitch circle.

$$p_c = \frac{\pi D}{N}$$

where $N =$ number of teeth in the gear.

Diametral Pitch, p_d. The number of teeth per inch of pitch diameter. A pair of gears must have the same diametral pitch (or the same

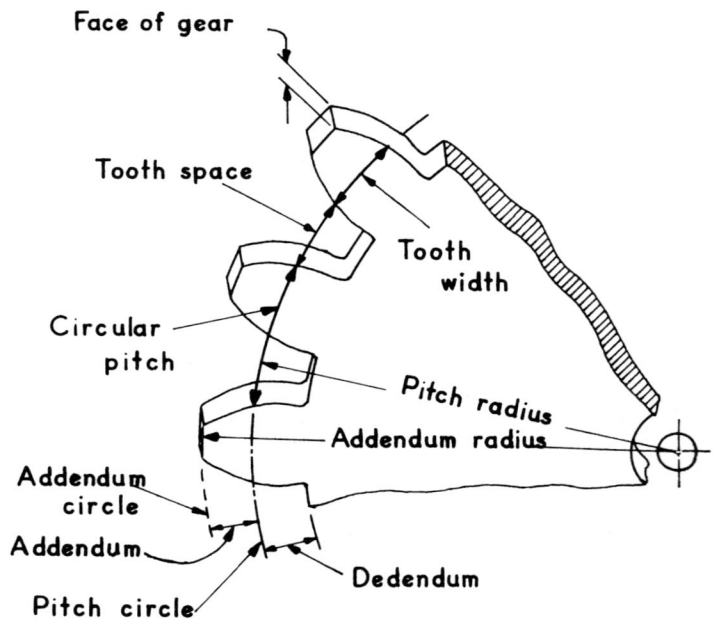

FIG. 5-9   TERMINOLOGY FOR GEARS.

circular pitch) in order to mesh.

$$p_d = \frac{N}{D}$$

Also

$$p_c \times p_d = \frac{\pi D}{N} \times \frac{N}{D} = \pi$$

Some sizes of gear teeth of different diametral pitches are shown in Fig. 5-10. Note that a large diametral pitch number indicates a small tooth; diametral pitch is an indicator of tooth size.

Module, m. Reciprocal of the diametral pitch.

$$m = \frac{1}{p_d}$$

Center Distance. The sum of the radii of the pitch circles of the mating gears.

$$C = \frac{D_1 + D_2}{2} = \frac{1}{2}\left(\frac{N_1}{p_d} + \frac{N_2}{p_d}\right) = \frac{N_1 + N_2}{2p_d}$$

Addendum, A. The radial distance from the pitch radius to the outside radius of the tooth.

$$A = 1/p_d$$

Dedendum. The radial distance from the pitch radius to the bottom of the tooth space.

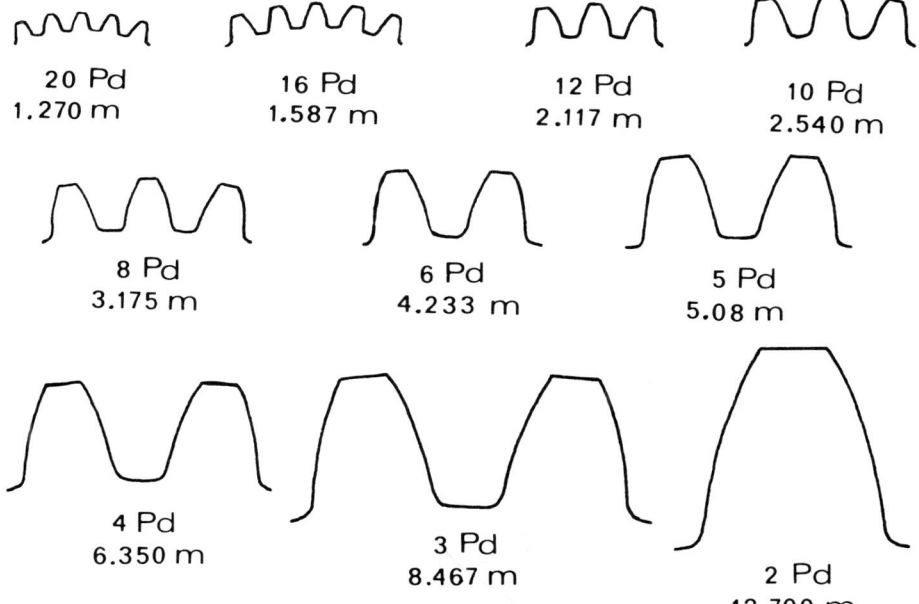

20 Pd
1.270 m

16 Pd
1.587 m

12 Pd
2.117 m

10 Pd
2.540 m

8 Pd
3.175 m

6 Pd
4.233 m

5 Pd
5.08 m

4 Pd
6.350 m

3 Pd
8.467 m

2 Pd
12.700 m

FIG. 5-10   DIAMETRAL PITCHES.

Whole Depth. The total height of the tooth: sum of addendum and dedendum.

Clearance. The space between the outside circle of one gear and the root circle of the mating gear.

Face and Flank of Tooth. The face is the contacting surface of the tooth from the pitch circle to the outside diameter or addendum circle. Similarly, the flank is the contacting surface from the pitch circle to the dedendum circle.

In North America gears have been designated by diametral pitch. But if gears are metric they are based on the module rather than diametral pitch. This change to module presents no problem, since the module is the reciprocal of diametral pitch.

$$m = 1/p_d = D/N \text{ millimeters.}$$

For diametral pitch, p_d is replaced by $25.4/m$.

The following formulas based on m are reasonably obvious and therefore need not be discussed.

$$D = mN \text{ mm.}$$
$$N = D/m$$
$$A = 1/p_d = m \text{ (millimeters)}$$
$$C = \tfrac{1}{2}m(N_1 + N_2) \text{ mm}$$
$$\text{Circular pitch} = m\pi = \frac{\pi D}{N}$$
$$\text{O.D. of gear} = D + 2m = m(N + 2)$$

Example. A pair of mating metric gears of module 1.25 have 28 and 38 teeth. Find the center distance, the pitch diameters and the addendum.

Solution.

$$C = \tfrac{1}{2}m(N_1 + N_2)$$
$$= \tfrac{1}{2}(1.25)(28 + 38) = 41.25 \text{ mm}$$
$$D = mN$$
$$= 1.25 \times 28 \text{ and } 1.25 \times 38, \text{ or } 35 \text{ and } 47.5 \text{ mm}$$
$$A = m = 1.25 \text{ mm}$$

Table 4-1 compares gears of various diametral pitches and standard modules.

5.5. TOOTH CONTACT.

If the tooth width of one gear is slightly smaller than the space width on the mating gear, as is usual to avoid binding, the gear drive still functions effectively. This difference, Fig. 5-11, is called *back-*

TABLE 4-1 STANDARD DIAMETRAL PITCHES AND MODULES
module 1 and larger sizes only

Diametral Pitch	Circular Pitch, Inches	Module, Milli-meters	Diametral Pitch	Circular Pitch, Inches	Module, Milli-meters	Diametral Pitch	Circular Pitch, Inches	Module, Milli-meters
25.400	.1237	1	3.590	$\frac{7}{8}$	7.074	1.270	2.4737	20
20.320	.1546	1.25	$3\frac{1}{2}$	.8976	7.257	1.257	$2\frac{1}{2}$	20.213
20	.1570	1.270	3.175	.9895	8	$1\frac{1}{4}$	2.5133	20.320
16.933	.1855	1.5	3.142	1	8.085	1.142	$2\frac{3}{4}$	22.234
16	.1963	1.587	3	1.0472	8.467	1.047	3	24.255
12.700	.2474	2	2.792	$1\frac{1}{8}$	9.096	1.016	3.0921	25
12	.2618	2.117	2.540	1.2368	10	1	3.1416	25.400
10.160	.3092	2.5	2.513	$1\frac{1}{4}$	10.106	.967	$3\frac{1}{4}$	26.276
10	.3142	2.540	$2\frac{1}{2}$	1.2566	10.160	.898	$3\frac{1}{2}$	28.298
8.467	.3711	3	2.285	$1\frac{3}{8}$	11.117	.838	$3\frac{3}{4}$	30.319
8	.3927	3.175	2.117	1.4842	12	.794	3.9579	32
6.350	.4947	4	2.094	$1\frac{1}{2}$	12.128	.785	4	32.340
6.2832	$\frac{1}{2}$	4.042	2	1.5708	12.700	$\frac{3}{4}$	4.1888	33.867
6	.5236	4.233	1.933	$1\frac{5}{8}$	13.138	.698	$4\frac{1}{2}$	36.383
5.080	.6184	5	1.795	$1\frac{3}{4}$	14.149	.635	4.9474	40
5.026	$\frac{5}{8}$	5.053	$1\frac{3}{4}$	1.7952	14.514	.628	5	40.425
5	.6283	5.080	1.587	1.9790	16	.524	6	48.510
4.233	.7421	6	1.571	2	16.170	.508	6.1842	50
4.189	$\frac{3}{4}$	6.064	$1\frac{1}{2}$	2.0944	16.933	$\frac{1}{2}$	6.2832	50.800
4	.7854	6.350	1.396	$2\frac{1}{4}$	18.191			

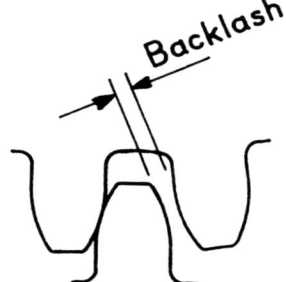

FIG. 5-11   BACKLASH.

lash. If the drive is reversed, the tooth of one gear loses contact with the other gear and lashes back to strike the tooth behind.

Contact between two mating gears as they rotate is shown in Fig. 5-12. Contact is first made at the corner of the face of the follower tooth (right hand side of the figure). The contact points are marked in series in the figure; note that contact proceeds on a straight diagonal line to the last point of contact on the corner of the driver tooth. The pitch point is the midpoint of this line of contact.

The tooth profile is an involute curve for standard gears, and Fig. 5-12 applies to involute teeth. A straight line of contact might not apply to other tooth forms.

The line of contact makes some angle ϕ to the common tangent to the pitch circles and is called the *pressure angle*. This pressure angle may be $14\frac{1}{2}°$ or $20°$ for power transmission gears, but is frequently $28°$ for gears used in gear pumps. The pressure angle is the angle between the common tangent to the base circles for the

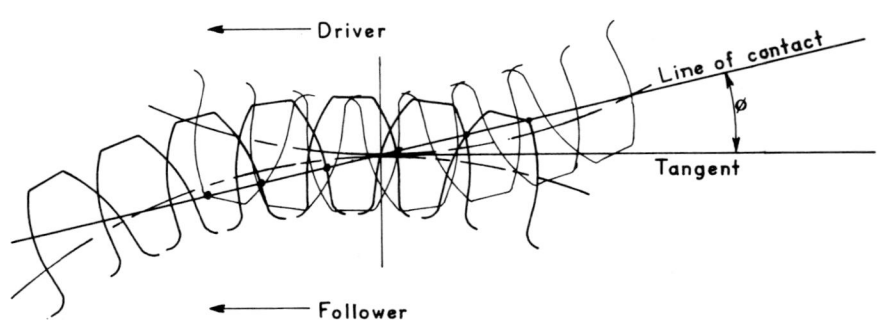

FIG.   5-12   SUCCESSIVE   POSITIONS   OF   ENGAGING
           TEETH.

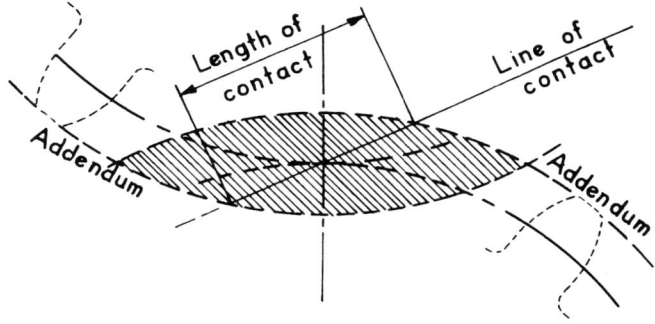

FIG. 5-13   LENGTH OF PATH OF CONTACT BETWEEN
TWO GEARS.

involutes and a perpendicular to the line of centers.

$$\text{Pressure angle} = \phi = \cos^{-1}\frac{\text{base circle dia}}{\text{pitch dia}}$$

Contact between two mating gears is possible only within an area defined by the overlapping addendum circles. This area is shown in Fig. 5-13. The line of contact extends to the limits of this shaded area. The length of the line of contact can be determined from the addendum circles and the pressure angle. This length is most conveniently found by graphical construction.

When a pinion (a pinion is the smaller of two mating gears) with a normal tooth form has fewer than a certain number of teeth, there is interference between the meshing teeth. This interference

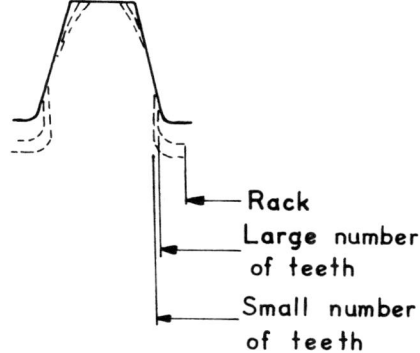

FIG.   5-14   EFFECT   OF   INTERFERENCE   ON   TOOTH
SHAPE.

occurs at the bottom of the pinion teeth, shown in Fig. 5-14. The minimum number of teeth that provides no interference or required undercutting of the teeth is 32 for a $14\frac{1}{2}°$ pressure angle and 18 for a 20° pressure angle. For a 20° pressure angle and stub teeth, the minimum number is 14 teeth. Undercutting is not desirable because the strength of the tooth is weakened.

If a gear must be keyed to a shaft, the minimum pitch diameter must be twice the shaft diameter. The face width of a gear is usually 3 to 4 times the circular pitch.

In order to mate, two gears must meet the following requirements:

1. Identical pitch

2. Identical pressure angle

3. Identical addendum and dedendum

Standard Gear Systems

	$14\frac{1}{2}°$ and 20° Full Depth (ASA B.6)	20° Stub (ASA B.6)	20° and 25° Full Depth Coarse Pitch 19.99 and Coarser (AGMA 201.02)	$14\frac{1}{2}°$, 20°, and 25 Full Depth, 20 Pitch and Finer (AGMA 207.05)
Addendum	$\dfrac{1.000}{P_d}$	$\dfrac{0.800}{P_d}$	$\dfrac{1.000}{P_d}$	$\dfrac{1.000}{P_d}$
Dedendum	$\dfrac{1.157}{P_d}$	$\dfrac{1.000}{P_d}$	$\dfrac{1.250}{P_d}$	$\dfrac{1.200}{P_d}+0.002$
Clearance	$\dfrac{0.157}{P_d}$	$\dfrac{0.200}{P_d}$	$\dfrac{0.250}{P_d}$	$\dfrac{0.200}{P_d}+0.002$

Example 1.   A 6 DP spur gear pinion with 18 teeth rotates at 1500 rpm driving a gear at 900 rpm. Find: (a) the number of teeth in the gear, (b) pitch diameter of the pinion, (c) pitch diameter of the gear, (d) the module, (e) O.D. of the pinion, (f) O.D. of the gear, and (g) center distance between shafts.

Solution.

a) $\dfrac{1500}{900} \times 18$ teeth $= 30$ teeth

b) $P_d = 6 = \dfrac{N}{D} = \dfrac{18}{D}$

$D = 3.00$ in.

c) $6 = \dfrac{30}{D}$ $D = 5.00$ in.

d) $M = \frac{1}{6}$

e) O.D. $=$ pitch diameter $+ 2$ addendums

$$A = \frac{1}{P_d} = \tfrac{1}{6}$$

$$\text{O.D.} = 3.00 + 2\left(\tfrac{1}{6}\right) = 3.333 \text{ in.}$$

f) O.D. $= 5.00 + 0.333 = 5.333$ in.

g) Center distance $=$ sum of pitch radii

$$= \frac{3.00 + 5.00}{2} = 4.00 \text{ in.}$$

Example 2. The driving gear is a mating pair of 20° full-depth spur gears that rotates at 3600 rpm. The driven gear must rotate at 1625 rpm as closely as possible. Center distance must be approximately 1.25 in., and fine-pitch gears of 32, 48, or 64 diametral pitch may be used. Select a suitable diametral pitch, pitch diameters, and number of teeth for an output speed closest to 1625 rpm from the three allowable diametral pitches.

Solution. The required speed ratio is $1625/3600 = 0.451$ closely.

Solving for driver pitch diameter we obtain:

$$D_1 = 2C\left(\frac{\omega_2}{\omega_1 + \omega_2}\right) = 2 \times 1.25\left(\frac{1625}{1625 + 3600}\right) = 0.7775 \text{ in.}$$

Then the approximate pitch diameter of the driven gear must be

$$2.50 - 0.777 = 1.723 \text{ in.}$$

The three diametral pitches are worked in the following tabulation:

p_d	$N_1 = D_1 p_d$	Whole N_1	$N_2 = 2CP_d - N_1$	N_2	N_2/N_1
32	.777×32=24.86	25	(2.5)32−25=55	55	0.454
48	.777×48=37.3	37	(2.5)48−37=83	83	0.446
64	.777×64=49.7	50	(2.5)64−50=110	110	0.454

Either 32 or 64 diametral pitch will give the speed ratio very closely. Using a p_d of 32, the pitch diameters are 25/32 and 55/32 or 0.781 and 1.72. The output speed will be 0.454×3600=1634 rpm.

5.6.   HELICAL GEARS.

Like a straight spur gear, a helical gear is cut from a cylindrical gear blank, but the teeth are helical, that is, if the gear had a sufficient axial length the teeth would appear as threads on a screw. Helical gears are preferred for high speeds, heavy loads, and low noise levels. The whole length of the tooth does not engage all at once but engages gradually, and this effect reduces noise and impact. While helical gears are suitable for mating to one another at 90°, at this angle there is only point contact between the mating gears, so that only low levels of power can be transmitted at 90°.

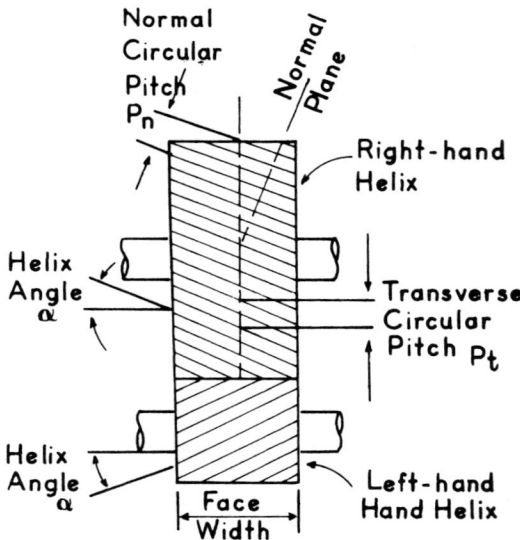

FIG. 5-15   HELICAL GEARS.

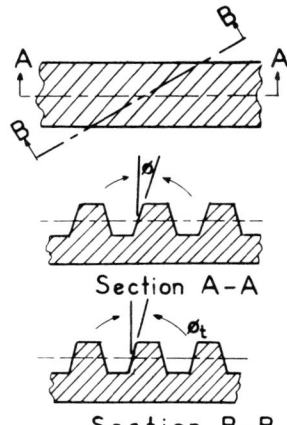

Section A-A

Section B-B

FIG. 5-16   MEASUREMENT OF PRESSURE ANGLE IN A
HELICAL GEAR.

The basic involute shape and the usual pressure angles apply to
helical gears.

The pressure angle ϕ in Fig. 5-16 is measured at the pitch circle.
It may be measured in either of two methods, which give two
different values for ϕ:

1. in a plane perpendicular to the axis of the gear

2. in a plane normal to the tooth.

The helix angle α is the angle to the axis of the gear, as shown
in Fig. 5-15. The hand of the helix must be established also. If two
mating helical gears are mounted on parallel shafts, one of the pair
must have a right-hand helix and the other a left-hand helix, as in
Fig. 5-15. If the gears are at 90° to each other, both gears must have
the same hand. The hand is designated in the usual way for screws: a
screw has a right-hand thread if when turned clockwise it advances
into a nut.

The helical advance of one tooth should preferably be at least
15% greater than the circular pitch. This ensures that at least one
pair of teeth will be in contact at the pitch point.

A disadvantage of helical gears is an axial thrust, which must
be resisted by a suitable thrust bearing, usually a ball or roller
bearing type. This thrust can be eliminated by the use of herring-
bone gears, which are equivalent to two side-by-side helical gears of
opposite hand. The direction of thrust depends on the hand and
direction of rotation (see Fig. 5-17).

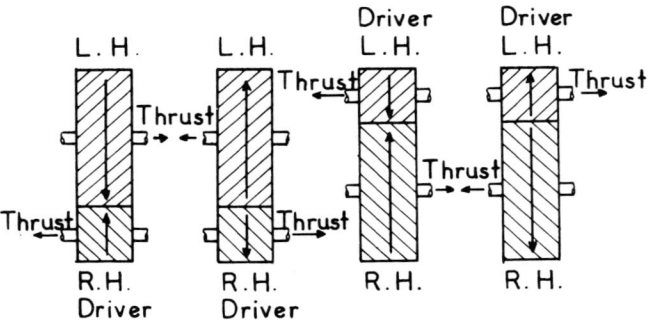

FIG. 5-17   DIRECTION OF THRUST IN HELICAL GEARS.

As with the pressure angle, the circular pitch p for helical gears may be measured in the traverse plane normal to the axis of the gear (p_t) or normal to the tooth (p_n), and

$$p_n = p_t \cos \alpha$$

Diametral pitches are related to the circular pitch in the same way as they are for spur gears:

$$\text{normal diametral pitch} = p_d^n = \frac{\pi}{p_n}$$

$$\text{transverse diametral pitch} = p_d^t = \frac{\pi}{p_t}$$

$$p_d{}^t = p_d^n \cos \alpha \qquad \text{also } p_d^n = \frac{N}{D \cos \alpha}$$

The center distance for helical gears

$$= C = \frac{D_1}{2} + \frac{D_2}{2} = \frac{N_1 + N_2}{2p_d{}^t} = \frac{N_1 + N_2}{2p_d^n \cos \alpha}$$

where subscripts 1 and 2 refer to gear 1 and gear 2. The speed ratio for a pair of helical gears is N_1/N_2.

Example.   A pair of helical gears, normal diametral pitch 4 and center distance 12.963 in., produce a speed reduction of 6:1. The pinion has 12 teeth. Determine pitch diameters, number of teeth, and helix angle.

Solution.   The large gear has 72 teeth. The pitch diameter of the small gear $= 2 \times 12.963/(1+6) = 3.704$ in. and the pitch diameter of the large gear $= 6 \times 3.704 = 22.224$ in. For the helix angle

$$\cos \alpha = \frac{N_1 + N_2}{2p_d^n c}$$

$$= \frac{12 + 72}{2 \times 4 \times 12.963}$$

$$= 0.81$$

$$\alpha = 36°$$

5.7.   WORM DRIVES.

A worm drive, consisting of a worm and gear (Fig. 5-18), has some resemblance to crossed helical gears but has a much greater power capacity. Worm drives are used to obtain large speed reductions in crossed shafts. The worm may have one, two, three, or four threads (single, double, triple, or quadruple worm). A single worm gives a speed reduction equal to the number of teeth in the gear; a double worm provides a speed reduction of half the number of teeth in the gear. Axial pitch of the worm must equal the circular pitch of the gear.

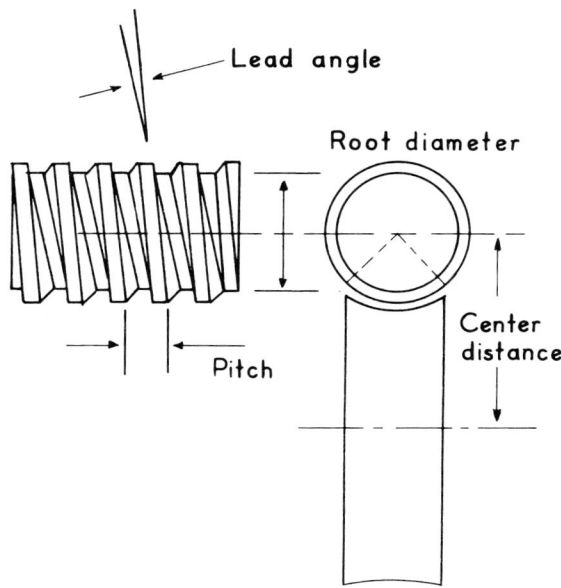

FIG. 5-18   TERMINOLOGY FOR A WORM DRIVE.

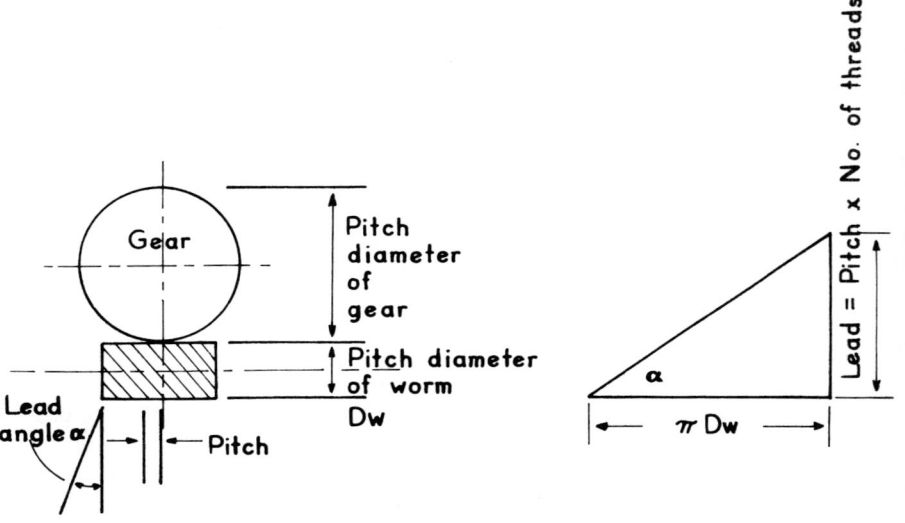

FIG. 5-19   LEAD ANGLE RELATIONSHIPS.

The *lead* of the worm is the tooth pitch times the number of threads. The lead angle α of the worm is equal to the helix angle of the gear.

$$\tan \alpha = \frac{\text{lead}}{\pi \,(\text{pitch dia})} \qquad \text{(Fig. 5-19)}$$

If the lead angle is small, as in a single-thread worm, the drive is self-locking, that is, the gear cannot drive the worm. Self-locking worms do not require brakes. However multithread worms are preferred: there is less sliding and therefore less wear and higher efficiency. Because the worm and gear have sliding contact, they are made of different metals to prevent seizing. Usually the worm is a hardened steel and the gear a bronze.

The American Gear Manufacturers Association (AGMA) recommends the following design formulas:

pitch diameter of worm = $C/2.2 = 3p_c$ approximately

where C = center distance between the two shafts.

Face width of gear = $0.73 \times$ worm pitch diameter approximately.

$$\text{Axial length of worm} = \left(4.5 + \frac{N_G}{50}\right)p_c$$

where N_G = number of gear teeth. The number of worm threads plus number of gear teeth $\geqslant 40$.

5.8.   BEVEL GEARS.

Bevel gears (Fig. 5-20) resemble a friction cone drive and thus connect shafts with intersecting axes, usually at 90°. The point of intersection of the two shafts is usually the apex of both pitch cones. The tooth length does not exceed one-third of the full length of the cone. Terminology of bevel gears is given in Fig. 5-20. Dimensions such as pitch diameter, diametral pitch, addendum, and dedendum, are measured at the large diameter of the bevel gear.

A pair of bevel gears with equal numbers of teeth and with axes at 90° is called a pair of miter gears.

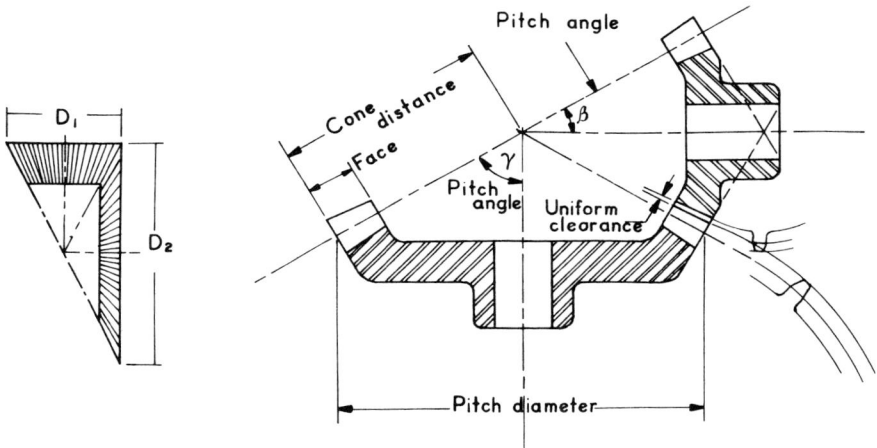

FIG. 5-20   TERMINOLOGY FOR BEVEL GEARS.

FIG. 5-21   SPIRAL BEVEL GEARS.

The tooth action of *spiral bevel gears* (Fig. 5-21) is similar to that of helical gears, in that tooth contact begins gradually. The tooth curve is not a spiral or helix (spiral bevel gears should be termed helical bevel gears) but a circular arc.

5.9.   HYPOID GEARS.

Hypoid gears connect nonparallel, nonintersecting shafts (see Fig. 5-22). There is sliding contact between hypoid gears. The two shafts can be extended beyond the gears without interfering with each other. This type of gear was originally developed for automobile rear ends, because with it the drive shaft could be lowered and therefore also the floor of the vehicle.

The pitch surfaces of hypoid gears are hyperboloids of revolution. Fig. 5-23 shows two hyperboloids. Hyperboloid 1 is produced by rotating line A—A about axis B—B with distance R_1 and angle ϕ constant. Hyperboloid 2 is generated by rotating A—A about axis C—C with distance R_2 and angle θ constant. If the hyperboloids make contact along A—A, then R_1 and R_2 must be proportional to the

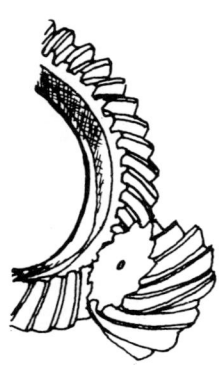

FIG. 5-22   HYPOID GEARS.

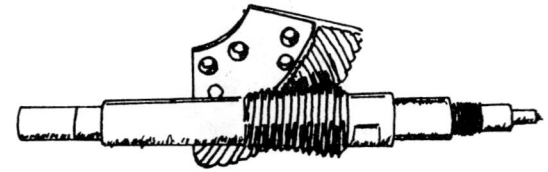

FIG. 5-23   ROLLING HYPERBOLOIDS.

tangents of the angles made by line A—A with the axes B and C:

$$\frac{R_1}{R_2} = \frac{\tan\phi}{\tan\theta}$$

The pitch surfaces of hypoid gears are portions of these hyperboloids.

5.10.   SPIROID GEARS.

A spiroid gear drive (Fig. 5-24) has a tapered worm. This type of drive has great shock resistance and backlash control. Several teeth are always in contact. Backlash is removed simply by adjusting either gear or worm axially. Spiroid gears are found in many portable tools such as grinders, drills, and hedge trimmers and in electric can openers.

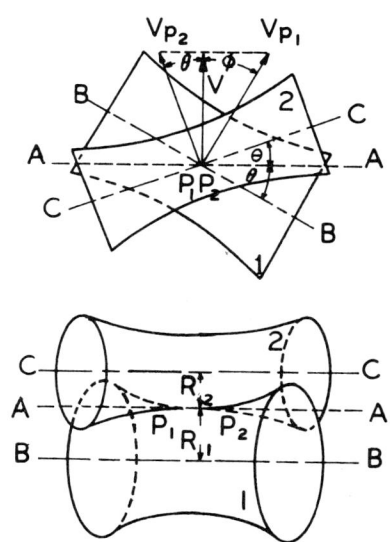

FIG. 5-24   SPIROID GEARS.

PROBLEMS

1. It would seem to be a fundamental law of the universe that two mating gears must rotate in opposite directions. Consider the two-tooth gear system of the figure in (a). With this basic unit the driver, of course, cannot drive the driven gear. But now laminate three pairs of these two-tooth gears as in (b) of the figure. The arrangement makes it possible to drive the other gear.

    If the triple-gear driver rotates in one direction, does the driven gear rotate in the opposite or the same direction? A hasty answer is ill advised.

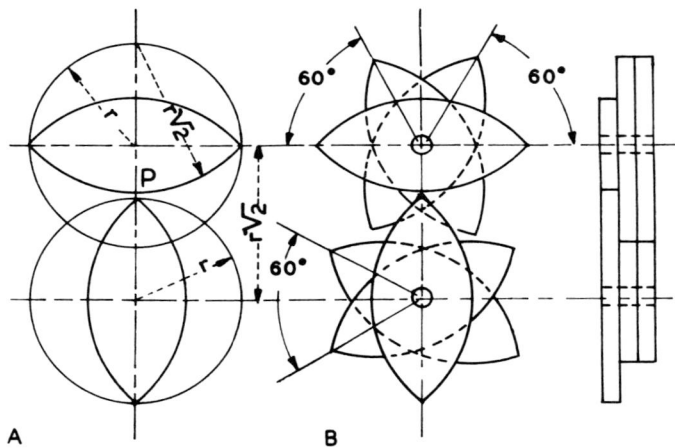

FIG. P5-1 (A) TWO-TOOTH GEAR.
　　　　　　(B) TRIPLE MOUNTING FOR CONSTRAINED MOTION.

2. A pair of mating 20° full-depth spur gears, 5.0 in. and 14.0 in. in pitch diameter, are selected from a catalogue. Determine the length of contact using the method of Fig. 5-13, assuming a diametral pitch of 5.

3. What is the pitch diameter of the following spur gears?
    a) 8 pitch, 16 teeth
    b) $1\frac{1}{2}$ pitch, 42 teeth
    c) 12 pitch, 16 teeth
    d) 5 pitch, 28 teeth.

4. What is the outside diameter for each of the gears of problem 3?

5. A small gear of 12 diametral pitch has 11 teeth. For this gear determine:
   a) Pitch diameter
   b) Addendum
   c) Outside diameter
   d) Dedendum, specification ASA B.6

6. What is the difference between
   a) Pitch diameter and diametral pitch
   b) Diametral pitch and circular pitch?

7. A full-depth spur gear has an O.D. of 3.250 in., a pitch diameter of 3.000 in., and 24 teeth. Find
   a) Addendum
   b) Diametral pitch
   c) Whole depth of tooth.

8. For a 20° full-depth spur gear, AGMA 201.02, 2 pitch and 90 teeth, determine the pitch diameter and outside diameter.

9. A 10 P_d spur gear pinion with 36 teeth rotates at 1800 rpm driving a gear at 720 rpm approximately. Find
   a) Number of teeth on the gear
   b) Center distance

10. The speed ratio for a pair of spur gears reduces the output gear speed to one-third that of the driving gear. The center distance is 1.40 in. Determine the pitch diameters.

11. A 36-tooth spur gear drives an internal gear with 84 teeth. Diametral pitch is 12. What is the center distance?

12. The driving gear for a pair of mating 20° full-depth gears rotates at 900 rpm and the driven gear must rotate at 400 rpm. A center distance of 3.50 in. approximately is required. Using any pitch of 10 to 14 inclusive and a minimum tooth number of 20, determine a suitable pair of gears and state the output rpm given by your choice of gears.

13. The driver pinion of a pair of mating 20° full-depth gears is to rotate at 1800 rpm and the driven gear is to rotate at 585 rpm approximately. A center distance of 2.40 in. approximately is required. Find a suitable pair of gears of 14, 16, or 18 diametral pitch and state the output rpm.

14. Two mating spur gears must have a center distance not exceeding 4.0 in. and a minimum number of teeth of 20. The output gear must rotate at 4.5 times the speed of the input gear. Pitches of 5, 6, 8, 10, or 12 are acceptable. Decide tooth numbers, diametral pitch, and center distance.

15. The figure shows a handling device which rotates an arm through 135°. The four-bar linkage drives the larger spur gear which then drives the upper pinion to which the pivoting arm is attached. The crank of the four-bar linkage rotates at a constant angular velocity. Use a gear ratio of 2.5:1 and any suitable diametral pitch of 20 to 30 inclusive to design the mechanism completely.

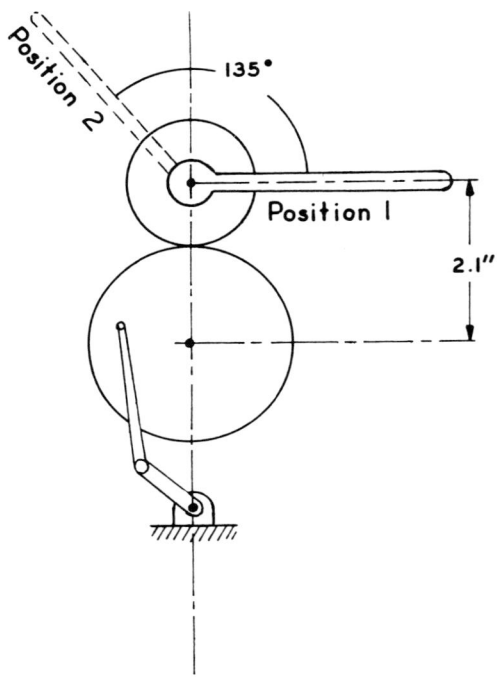

Position 2

135°

Position I

2.1"

FIG. P5-15   PACKAGE HANDLING MECHANISM.

16. An electric motor speed of 1160 rpm is to be reduced by a gear ratio of 7:1 using a pair of helical gears on a center distance of 21.75 in. with a normal diametral pitch of 4. The pinion has 20 teeth. Determine the number of teeth, pitch diameters, and helix angle.

17. A helical gear with a helix angle of 15° and a normal diametral pitch of 16 has 50 teeth. Determine
    a) Pitch diameter
    b) Transverse diametral pitch

18. A pair of helical gears has a center distance of 3.1 in. and a helix angle of 14°30′. The gears have 20 and 40 teeth. Find the normal and transverse diametral pitches.

19. Design a pair of helical gears for a speed reduction of 7:1 with a center distance of 2.00 in. Assume a helix angle of 25° and a pinion with 11 teeth, and use a normal diametral pitch within the range of 20 to 30 inclusive.

20. If in Problem 19 the larger gear must be overhung and supported by a bearing only on the left-hand side of the gear, decide the hand of both gears.

21. A worm drive is to produce a speed reduction of 50:1 using a double-thread worm, and a diametral pitch of 48 for the gear. Determine:
    a) The exact center distance
    b) The helix angle of the gear
    c) The lead angle of the worm
    d) The pitch diameters

22. A triple-thread worm has a pitch diameter of 4.00 in. and an axial pitch of 0.875 in. What is the lead angle?

23. A pair of mating straight-tooth bevel gears at 90° to each other have a speed ratio of 4:3. The pinion has a pitch diameter of 6.00 in. and a face width of 1.500 in., rotating at 540 rpm. Diametral pitch is 5, with a 20° pressure angle. Determine the pitch diameter of the larger bevel gear and the diameter of the large and the small end of this gear.

24. A pair of elliptical gears both have major axes 3.5 cm long and a minor axis 2.0 cm long. If the driving ellipse rotates at 100 rpm, graph the rotational speed of the driven ellipse through one revolution. From the graph of angular velocity, draw the graph of angular acceleration.

25. The figure shows a two-lobed blower rotor rotating about an axis at *B*. The mating rotor rotates about an axis at *A*. The two rotors are geared together and rotate at the same angular velocity but in opposite directions. Graphically design the mating rotor.

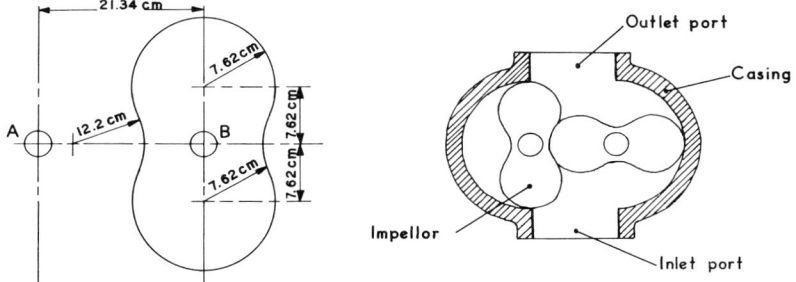

FIG. P5-25 CYCLOIDAL CONJUGATE LOBES FOR A ROOTS-TYPE BLOWER.

26. The figure shows the general shape of the mating rotors in an SRM air compressor. Graphically design the profile of the mating rotor and state whether or not the two rotors provide conjugate action.

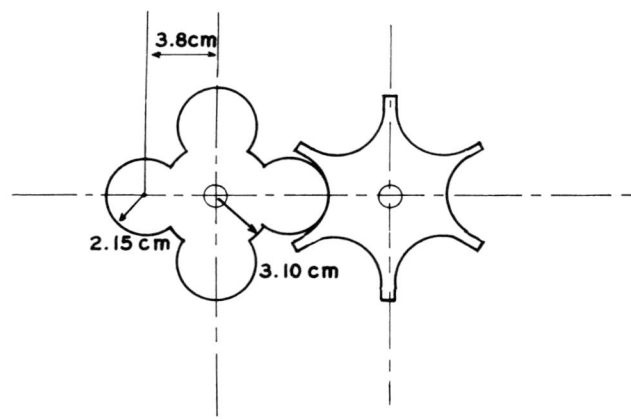

FIG. P5-26   MULTI-LOBED AIR BLOWER.

27. An elliptical gear with the dimensions given in Problem 24 mates with two gears, one being a twin elliptical gear and the other being a suitably profiled gear to give conjugate action with the elliptical gear and a 1:1 velocity ratio. Design the profiled gear.

28. A spur gear of module 4.233 mm and 18 teeth rotates at 1500 rpm driving a gear at 900 rpm. Find:
    a)   pitch diameter of the pinion
    b)   number of teeth in the gear
    c)   pitch diameter of the gear
    d)   diametral pitch
    e)   O.D. of pinion
    f)   center distance between shafts

29. A spur gear with 32 teeth and module 5.0 rotates at 1800 rpm driving a gear at 750 rpm. Find
    a)   number of teeth in the gear
    b)   center distance.

chapter six

Gear Trains

A power transmission system consisting only of gears is called a gear train. Chain drives are often used in association with gears and are considered to be a part of the train. Speed ratios are determined by the numbers of teeth in the units of the train, whether gears or sprockets.

Gear trains are classified as ordinary or simple, compound, reverted, or epicyclic (planetary). At least one of the gear axes must rotate about another axis in the case of the planetary train. For other gear train, all gear axes are fixed in position.

6.1. SIMPLE AND COMPOUND GEAR TRAINS.

In the ordinary or simple gear train, for which Fig. 6-1 is typical, each shaft carries only one gear. The pitch line velocities are the same for all gears. If gear A is taken to be the driving gear, then speed ratios are

$$\frac{\omega_B}{\omega_A} = \frac{N_A}{N_B} \qquad \frac{\omega_C}{\omega_B} = \frac{N_B}{N_C} \qquad \frac{\omega_D}{\omega_C} = \frac{N_C}{N_D}$$

and
$$\frac{\omega_D}{\omega_A} = \left(\frac{\omega_B}{\omega_A}\right)\left(\frac{\omega_C}{\omega_B}\right)\left(\frac{\omega_D}{\omega_C}\right) = \frac{N_A}{N_B}\frac{N_B}{N_C}\frac{N_C}{N_D} = \frac{N_A}{N_D}$$

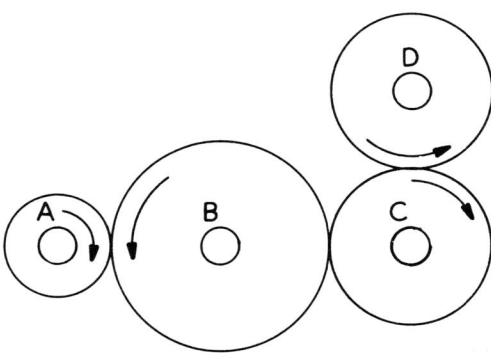

FIG. 6-1 SIMPLE GEAR TRAIN WITH FOUR GEARS; *A*
AND *D* ROTATE IN OPPOSITE DIRECTIONS.

The intermediate gears between driver and output gear do not
influence the overall speed ratio.

The simple gear train is used when a large center distance
between input and output must be spanned by the use of intermediate gears, to reverse the direction of the output shaft or to take
off power from intermediate gears. The two crankshafts of opposed-
piston diesel engines are connected by such a train.

In the compound gear train at least one shaft must carry two
gears. Consider the compound train of Fig. 6-2. The speed ratios are

$$\frac{\omega_B}{\omega_A} = \frac{N_A}{N_B} \quad \frac{\omega_D}{\omega_C} = \frac{N_C}{N_D} \quad \frac{\omega_F}{\omega_E} = \frac{N_E}{N_F}$$

The overall speed ratio $= \dfrac{\omega_F}{\omega_A} = \dfrac{\omega_B}{\omega_A} \times \dfrac{\omega_D}{\omega_C} \times \dfrac{\omega_F}{\omega_E}$

where $\qquad \omega_B = \omega_C \qquad$ and $\qquad \omega_D = \omega_E$

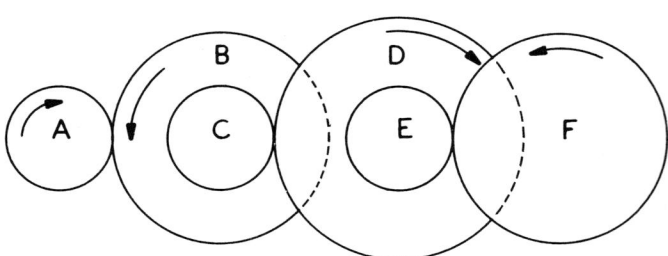

FIG. 6-2 COMPOUND GEAR TRAIN. GEAR *F* ROTATES
OPPOSITE TO GEAR *A*.

Substituting tooth numbers:

$$\frac{\omega_F}{\omega_A} = \frac{N_A}{N_B} \frac{N_C}{N_D} \frac{N_E}{N_F}$$

All the gears influence the overall speed ratio. Note that the numerator is the product of all the driver teeth ($N_A \times N_C \times N_E$) and the denominator is the product of all the driven teeth ($N_B \times N_D \times N_F$). The same relations hold for belt and chain drives using pitch diameters.

6.2.   DESIGN OF A COMPOUND GEAR TRAIN.

The design of a compound gear train is most easily understood from an example rather than a statement of procedure.

Suppose a compound gear train must provide a speed reduction of 1:15. Standard 20° gears of diametral pitch 24 are selected (this is a fine pitch). A maximum pitch diameter of 3 in. is desirable, and to prevent undercutting of pinions, the minimum number of teeth must be 18. All tooth numbers are available up to 25 teeth and all even numbers of teeth up to 100.

The first matter to be determined is the number of pairs of gears that are needed. Since a pitch diameter not greater than 3 in. is desired, the largest number of teeth desirable is 72 (3×24). The smallest allowable number of teeth is 18. The maximum reduction ratio then is 18/72 or 1:4. Two such reductions will give a speed reduction of 1:16. Therefore two pairs of gears will provide the required ratio of 1:15.

First consider two equal speed reductions:

$$\frac{\omega_{out}}{\omega_{in}} = \frac{1}{15} = \frac{1}{3.88} \times \frac{1}{3.88} \qquad \text{where} \qquad 3.88^2 = 15$$

and each reduction is 1:3.88. But 3.88 cannot be used. Change one of the 3.88's to some number close to 3.88 but which makes a reasonably simple fraction. While it is not the best choice, assume for illustration purposes that 3.875 (31/8) is selected. This fraction alters the speed ratios to

$$\frac{1}{15} = \frac{1}{\dfrac{31}{8}} \times \dfrac{\dfrac{31}{8}}{15} = \frac{8}{31} \left[\frac{31}{8 \times 15} \right] = \frac{8}{31} \left(\frac{31}{120} \right)$$

a reduction of 8/31. Then 31/120 (tooth numbers) will provide the speed reduction required. But 120 teeth is too large a number, and 31 teeth are not available. The first gear has only 8 teeth, and even if the first reduction is made 16/62, 16 teeth are still too few. The fraction 31/8 does not seem successful.

The choice of 1:4 for the first reduction seems a more natural choice:

$$\frac{1}{15} = \frac{1}{4} \times \frac{4}{15} = \frac{18}{72} \times \frac{20}{75}$$

Eighteen teeth is the choice for the first gear because 18 is the minimum number of teeth allowed. However, 75 teeth are slightly over the desired maximum, and also 75 teeth may not be available. Practical considerations may dictate acceptance of ratios $18/72 \times 20/74$ or $18/72 \times 20/76$. Normally such small modifications to a desired speed ratio are of no consequence.

For a second example, suppose a gear reducer must reduce an electric motor speed of 1750 rpm to 200 rpm. No gear is to have fewer than 18 teeth or more than 50. As in the previous example, odd numbers of teeth past 25 are assumed to be available only on special order.

The speed ratio $= 1750/200 = 8.75$

Using the smallest and largest allowable gears, 20 and 50 teeth, maximum speed reduction with one pair of gears is 50/20 or 2.5. Two such pairs will give a reduction of 6.25. Hence three pairs of gears are needed.

Since three combinations of gears are required, first find the cube root of 8.75, which is 2.06. It seems reasonable to try a 2:1 ratio for the first pair of gears, and perhaps for the second pair as well. A 2:1 ratio also ensures an even number of teeth for the larger gear of the pair.

$$2.0 \times 2.0 \times 2.1875 = 8.75, \quad \text{and} \quad 2.1875 = 35/16.$$

For the last ratio, 35/16, increase the tooth numbers by 25%: $35 \times 1.25/16 \times 1.25 = 43\frac{3}{4}/20$, adjusted to 44/20. For the first two ratios, use 20 and 40 teeth, or 22 and 44 teeth. There may be cost savings in using 3 gears of 44 teeth in the drive.

In these examples the matter of direction of rotation of the output shaft has been ignored, as it often can be. In the second example the gear train is powered by an electric motor; the direction of rotation can best be taken care of in the electric motor rather than in the drive.

Note that the same diametral pitch is not necessarily required for all pairs of gears in the train. As the speed drops, the torque increases, and it may be desirable to reduce the diametral pitch on slower shafts to gain a larger and stronger tooth.

Sometimes a very large speed reduction is required, such as 700:1. Such large reductions require worm drives, although epicyclic drives are occasionally used. A single worm and gear can reduce speed by as much as 60:1; larger reductions are too inefficient even though they are possible mechanically. A 700:1 reduction can be made by two worm reductions or by a standard gear train with a worm drive. A multiple-thread worm is preferred because of less friction and wear.

6.3. DESIGN OF A REVERTED GEAR TRAIN.

In a reverted gear train, both the input and the output shafts have the same line as the axis. See Fig. 6-3. The center distance must be the same for both pairs of gears.

If C = center distance, then

$$2C = D_A + D_B = D_C + D_D$$

But since

$$D = \frac{N}{p_d}$$

$$2C = \frac{N_A}{p_1} + \frac{N_B}{p_1} = \frac{N_C}{p_2} + \frac{N_D}{p_2}$$

where p_1 and p_2 are two different diametral pitches. If a single diametral pitch is used, then

$$2Cp_d = N_A + N_B = N_C + N_D$$

For a single diametral pitch the same total number of teeth must be used in each pair of gears.

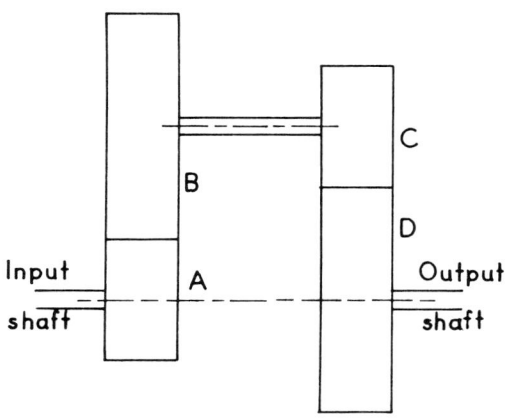

FIG. 6-3 FOUR GEARS IN A REVERTED GEAR TRAIN.

The design of a reverted gear train presents somewhat the same problem as the design of a compound gear train, with the added constraint of a fixed center distance. Again, the procedure is best understood from an example. The following example will use the same diametral pitch for all gears.

A reverted gear train is to provide a speed reduction of 30:1. As in the previous examples, gears with an odd number of teeth above 25 are specials and, if possible, should not be selected. No gear may have fewer than 18 teeth, and the maximum reduction per pair of gears is decided as 6:1.

Then two reductions will provide a 30:1 ratio. The square root of 30 is 5.48. Rather than using fractional ratios, it seems preferable to try reductions of 6:1 and 5:1.

$$\frac{1}{30} = \frac{1}{6} \times \frac{1}{5}$$
$$= \frac{18}{108} \times \frac{18}{90} \text{ using the minimum number of teeth.}$$

However the tooth numbers in each pair are not equal:

$$18 + 108 = 126 \text{ and } 18 + 90 = 108.$$

Suppose the ratios $1/6 \times 1/5$ were actual tooth numbers. To provide equal tooth numbers, $1/6$ would have to be multiplied by $6/6$ (the sum of $1+5$) and $1/5$ by $7/7$ (the sum of $1+6$).

$$\left(\frac{1}{6}\frac{6}{6}\right) \times \left(\frac{1}{5}\frac{7}{7}\right) = \frac{6}{36} \times \frac{7}{35}$$

and numbers of teeth are equal. To provide the required minimum of 18 teeth

$$\left(\frac{6}{36}\frac{3}{3}\right) \times \left(\frac{7}{35}\frac{3}{3}\right) = \frac{18}{108} \times \frac{21}{105}$$

There are 126 teeth in each pair. In a second example, the speed ratio is to be a reduction of 6:1. Minimum and maximum number of teeth in any gear are 18 and 96, and a uniform diametral pitch will be used.

The maximum reduction in a single pair of gears is $18/96$ or $3/16$, and in two pairs $9/64$. A ratio of 6:1 falls between these two limits, so two reductions are sufficient. Since the square root of 6 is 2.45, try a 2.5 reduction in the first ratio:

$$\frac{1}{6} = \frac{1}{2.5}\frac{2.5}{6}$$

But

$$1 + 2.5 \neq 2.5 + 6$$
$$\frac{1}{2.5} \times \frac{2.5}{6} = \frac{2}{5} \times \frac{5}{12}$$

with sums of 7 and 17. Multiply the first fraction by 17/17 and the second fraction by 7/7:

$$\left(\frac{2}{5}\frac{17}{17}\right) \times \left(\frac{5}{12}\frac{7}{7}\right) = \frac{34}{85} \times \frac{35}{84}$$

The tooth numbers are equal, but 85 and 35 teeth are not preferred numbers. Adjustment of tooth numbers to $34/84 \times 34/84$ will give a close approximation to 6:1, but this case is solved more easily by choosing reductions of $1/2 \times 1/3$, with sums of digits in the two ratios of 3 and 4:

$$\frac{1}{2} \times \frac{1}{3} = \frac{1}{2}\frac{4}{4} \times \frac{1}{3}\frac{3}{3} = \frac{4}{8} \times \frac{3}{9}$$

$$= \frac{24}{48} \times \frac{18}{54} \text{ or } \frac{32}{64} \times \frac{24}{72} \text{ or } \frac{40}{80} \times \frac{30}{90}$$

If different diametral pitches are selected for each reduction in a reverted gear train, then the procedure is slightly more complex. As an example, suppose the previous case of a 6:1 reduction is to be designed using a diametral pitch of 6 for the first reduction and 5 for the second reduction (larger teeth for higher torque requirements).

$$\frac{N_1 + N_2}{6} = \frac{N_3 + N_4}{5}$$

Multiply by 30, the product of the two diametral pitches:

$$5N_1 + 5N_2 = 6N_3 + 6N_4$$

$$\frac{\omega_{\text{out}}}{\omega_{\text{in}}} = \frac{1}{6} = \frac{\text{product of tooth numbers of driving gears}}{\text{product of tooth numbers of driven gears}}$$

$$= \frac{N_1}{N_2} \times \frac{N_3}{N_4}$$

This speed ratio is not changed if N_1 and N_2 are multiplied by 5 and N_3 and N_4 are multiplied by 6:

$$\frac{N_1}{N_2} \times \frac{N_3}{N_4} = \frac{5N_1}{5N_2} \times \frac{6N_3}{6N_4}$$

Using the same reductions as before, that is, $\dfrac{N_1}{N_2} = \dfrac{1}{2}$, $\dfrac{N_3}{N_4} = \dfrac{1}{3}$,

then

$$\frac{5 \times 1}{5 \times 2} = \frac{6 \times 1}{6 \times 3}$$

But $5 + 10$ is not equal to $6 + 18$. $5 + 10 = 15$, $6 + 18 = 24$. The smallest number divisible by 15 and 24 is 120, and $120/15 = 8$, and $120/24 = 5$. For the first reduction multiply by 8:

$$\frac{8 \times 5 \times 1}{8 \times 5 \times 2}$$

and for the second reduction multiply by 5:

$$\frac{5\times6\times1}{5\times6\times3}$$

Now $\quad (8\times5\times1) + (8\times5\times2) = 40 + 80 = 120, \qquad$ and

$\qquad\quad (5\times6\times1) + (5\times6\times3) = 30 + 90 = 120 \qquad$ also.

Finally, to obtain tooth number ratios, factor out the coefficients 5 and 6 that were used to obtain these ratios:

$$\frac{5(8\times1)}{5(8\times2)} \times \frac{6(5\times1)}{6(5\times3)} = \frac{8}{16} \times \frac{5}{15}\left(= \frac{1}{6}\right)$$

To obtain tooth numbers not less than 18, mutliply by 4:

$$\frac{32}{64} \times \frac{20}{60} = \frac{1}{6}$$

This is a satisfactory solution. As a check, calculate center distances for both reductions:

$$D = \frac{N}{p_d}$$

$$\frac{96}{6} = 16 \qquad \text{also} \qquad \frac{80}{5} = 16$$

For a second example, suppose the previous example is reworked using a diametral pitch of 8 in the first reduction and 6 in the second. The overall reduction is $1/2\times1/3=1/6$ as before.

$$\frac{N_1 + N_2}{8} = \frac{N_3 + N_4}{6}$$

Multiply by 24, which is divisible by both 8 and 6:

$$3N_1 + 3N_2 = 4N_3 + 4N_4$$

But $\qquad \dfrac{\omega_{\text{out}}}{\omega_{\text{in}}} = \dfrac{N_1}{N_2}\dfrac{N_3}{N_4} = \dfrac{3N_1}{3N_2}\dfrac{4N_3}{4N_4}$

$$\frac{N_1}{N_2} = \frac{1}{2} \quad \frac{N_3}{N_4} = \frac{1}{3}$$

$$\frac{3\times1}{3\times2} \times \frac{4\times1}{4\times3} = \frac{1}{6}$$

The sum of numerator and denominator for each reduction is 9 and 16. The smallest number divisible by both 9 and 16 is 144. Multiply the first reduction by 16 and the second by 9 to obtain:

$$\frac{16\times3\times1}{16\times3\times2} \times \frac{9\times4\times1}{9\times4\times3}$$

The sums are $48+96$ and $36+108$, or 144 in both cases. Finally, factor out the coefficients 3 and 4 originally used:

$$\frac{3(16\times1)}{3(16\times2)} \times \frac{4(9\times1)}{4(9\times3)} = \frac{16}{32} \times \frac{9}{27}$$

Multiplying by 2 will give tooth numbers none less than 18:

$$\frac{32}{64} \times \frac{18}{54}$$

Center distances are the same for both reductions.

6.4.   THE PLANETARY OR EPICYCLIC GEAR TRAIN.

A simple type of planetary gear train is shown in Fig. 6-4. The unusual characteristic of this train is the rotating arm or planet carrier. If the sun gear is stationary and the arm rotates clockwise about the axis 0, then also the planet gear rotates clockwise about its center. There are three possible operating methods for such a gear train:

1. Sun gear stationary and arm rotating.

2. Planet gear stationary with both arm and sun gear rotating.

3. Arm stationary and both gears rotating. This third case converts the train to a standard gear train.

The determination of the speed ratio of an epicyclic train is a slightly more complex an analysis than that required for other

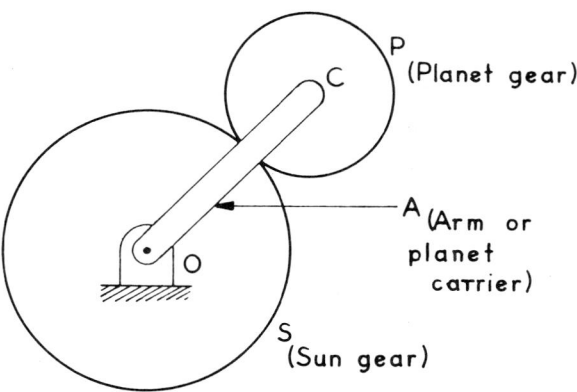

FIG. 6-4 BASIC PLANETARY GEAR TRAIN.

trains. For the train of Fig. 6-4, it is necessary to know the speeds of any two of the rotating members in order to find the speed of the third member.

Suppose the angular velocities of the arm A and the sun gear S are known, and for convenience, that both rotate clockwise. The sun gear rotates about axis O and the planet gear about axis Q. See Fig. 6-5.

$$v_Q = OQ\omega_A = (R_S + R_P)\omega_A \qquad \text{(Fig. 6-5)}$$
$$v_{P'} = R_S\omega_S$$

where P' is the pitch point.

$$\omega_P = \frac{v_{Q/P'}}{R_P}$$

and since $\quad v_{Q/_{P'}} = v_Q - v_{P'}$

$$\omega_P = \frac{v_Q - v_P}{R_P} \qquad \text{and} \qquad R_P\omega_P = v_Q - v_P$$

Using the equations for v_Q and v_P above in this equation for ω_P

$$\omega_P R_P = (R_S + R_P)\omega_A - R_S\omega_S$$
$$= R_S\omega_A + R_P\omega_A - R_S\omega_S$$
$$R_P(\omega_P - \omega_A) = R_S(\omega_A - \omega_S)$$
$$\frac{\omega_P - \omega_A}{\omega_A - \omega_S} = \frac{R_S}{R_P}$$

Next, expressing the pitch radii R_S and R_P in tooth numbers, from $p_d = N/D$:

$$R = \frac{N}{2p_d} \qquad R_S = \frac{n_S}{2p_d} \qquad \text{and} \qquad R_P = \frac{N_P}{2p_d}$$

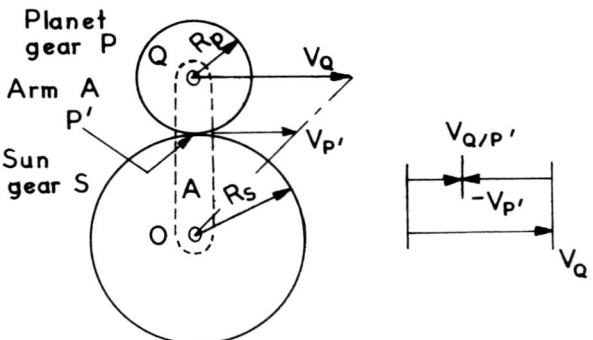

FIG. 6-5   VELOCITIES IN AN EPICYCLIC TRAIN.

and $\qquad \dfrac{\omega_P - \omega_A}{\omega_A - \omega_s} = \dfrac{N_s}{N_P}$

In the epicyclic train of Fig. 6-4 the ratio N_S/N_P is negative because both gears do not rotate in the same sense. Multiply both sides of the previous speed ratio equation by -1:

$$\frac{-\omega_p + \omega_A}{\omega_A - \omega_S} = -\frac{N_S}{N_p}$$

and then both numerator and denominator of the left hand side by -1:

$$\frac{\omega_P - \omega_A}{\omega_S - \omega A} = -\frac{N_S}{N_P}$$

How does this speed ratio compare with that of a standard pair of mating gears? For the standard pair, using the same subscripts as are used here for an epicyclic train

$$\frac{\omega_P}{\omega_S} = -\frac{N_S}{N_P}$$

For the epicyclic pair

$$\frac{\omega_{P/_A}}{\omega_{S/_A}} = -\frac{N_S}{N_P}$$

because the difference between two absolute angular velocities is a relative velocity: $\omega_{P/_A} = \omega_P - \omega_A$ and $\omega_{S/_A} = \omega_S - \omega_A$. Even the speed ratio for a standard pair is a ratio of relative angular velocities with respect to the gear box, the gear box being a kind of stationary arm in a planetary train.

Example 1.   In the train of Fig. 6-4, the sun gear has 60 teeth and rotates clockwise at 240 rpm. The planet gear has 20 teeth. The arm rotates counterclockwise at 100 rpm about 0. Determine the angular velocity of the planet gear.

$$\frac{\omega_P - \omega_A}{\omega_S - \omega_A} = -\frac{N_S}{N_P}$$

$$\frac{\omega_P - (-100)}{240 - (-100)} = -\frac{60}{20}$$

$$\frac{\omega_P + 100}{240 + 100} = -3$$

$$\omega_P = -920 \text{ rpm}$$

counterclockwise, since the clockwise motion of the sun gear is taken as plus.

Example 2. The epicyclic gear train is capable of very large speed reductions at a higher efficiency than a worm drive can provide, though it is possible to design an efficient epicyclic drive. Fig. 6-6 shows a reverted epicyclic train for a large speed reduction. Sun gear A, 80 teeth, is fastened to the gear box and does not rotate. Gear D, 81 teeth, is keyed to the output shaft. The gears B, 81 teeth, and C, 82 teeth, are keyed to an intermediate shaft carried on the arm. The arm is rotated by the input shaft. The similarity in tooth numbers is notable.

Solution. To determine the speed reduction of this train, let gear A be taken as the driver and gear D as the output gear, and take the speed of the input shaft and epicyclic arm as 1000 rpm. The train can be analyzed as two epicyclic trains in series or as an overall single train. Taking it first as two trains in series, for the first epicyclic train

$$\frac{\omega_P - \omega_{arm}}{\omega_S - \omega_{arm}} = -\frac{N_S}{N_P}$$

$$\frac{\omega_P - 1000}{0 - 1000} = -\frac{80}{81} = -0.98765$$

$$\omega_P = +12.35$$

In the second or output subtrain, gear C has the same speed as the planet gear B because both are keyed to the same shaft. For the second train

$$\frac{\omega_P - \omega_{arm}}{\omega_S - \omega_{arm}} = -\frac{N_S}{N_P}$$

$$\frac{12.35 - 1000}{\omega_S - 1000} = -\frac{81}{82} = -0.987805$$

$$\omega_S = 0.152 \text{ rpm}$$

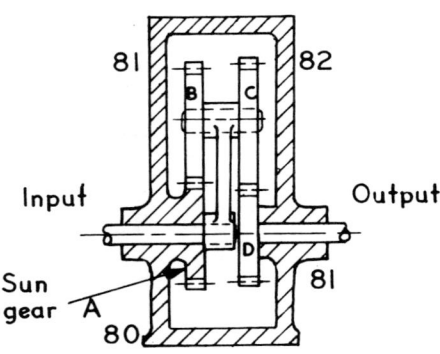

FIG. 6-6   REVERTED EPICYCLIC TRAIN.

The rotation of the input shaft was taken as positive, and since ω_S is positive, it rotates in the same sense as the input shaft.

To analyze the whole train as a unit, gear A is to be taken as the driver and D as the output gear.

$$\frac{\omega_D - \omega_{\text{arm}}}{\omega_A - \omega_{\text{arm}}} = \frac{N_A \times N_C}{N_B \times N_D} \text{ (products of input and output gears as for any gear train)}$$

$$\frac{\omega_D - 1000}{0 - 1000} = \frac{80 \times 82}{81 \times 81} = \frac{6560}{6561}$$

$$\omega_D = \frac{1000}{6561} = 0.152 \text{rpm}$$

Note that if gear A were able to rotate, gear D would rotate in the same sense as A, so that the input and output shafts would rotate in the same direction.

A ring gear is often used in epicyclic trains, as in Fig. 6-7. Two planet gears are shown. The use of one planet gear results in unbalance. Two or more planetary gears also increase torque capacity. This type of epicyclic train can be analyzed by the same methods as before. Suppose the input gear is the sun gear and the output gear is the ring gear. Then, as before for the general case,

$$\frac{\omega_{\text{out}}}{\omega_{\text{in}}} = \frac{N_{\text{in}}}{N_{\text{out}}}$$

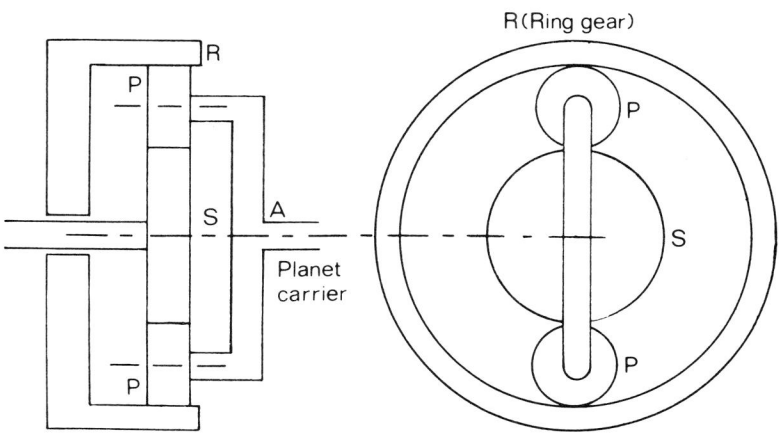

FIG. 6-7 EPICYCLIC TRAIN WITH RING GEAR. TWO PLANET GEARS ARE USED FOR INCREASED TORQUE CAPACITY AND DYNAMIC BALANCE.

But for the epicyclic train shown

$$\frac{\omega_R - \omega_A}{\omega_S - \omega_A} = -\frac{N_S}{N_R}$$

Example.   Determine the angular velocity of the arm of Fig 6-7 given the following information:

$N_S = 80$, $N_P = 20$, $N_R = 120$, $\omega_S = 1800$ rpm clockwise, $\omega_R = 0$

Solution.   The ring gear will be taken as the output gear. Although it does not rotate, that circumstance does not restrict this gear from being taken as the output gear.

$$\frac{\omega_R - \omega_A}{\omega_S - \omega_A} = -\frac{N_S}{N_R}$$

$$\frac{0 - \omega_A}{1800 - \omega_A} = -\frac{80}{120}$$

$$\omega_A = 720 \text{ rpm}$$

Since ω_A is positive, the arm rotates in the same sense as the sun gear.

Another equation often used in analyzing epicyclic gear trains is the following:

$$\omega_S = \left(1 + \frac{N_R}{N_S}\right)\omega_A - \frac{N_R}{N_S}\omega_R$$

Using this equation for the present example:

$$\omega_S = 1800 = \left(1 + \frac{120}{80}\right)\omega_A - \frac{120}{80}(0)$$

$$\omega_A = 720 \text{ rpm}$$

It should be noted that the following relationship must hold for such an epicyclic train:

$$N_S + 2N_P = N_R$$

Table 6-1 classifies the types of three-gear planetary trains consisting of a sun and planet gear (or gears), a planet carrier, and a ring gear. The table is explained in the following discussion. The equation for such a train is

$$\frac{\omega_R - \omega_A}{\omega_S - \omega_A} = -\frac{N_S}{N_R}$$

which expands to

$$N_R\omega_R + N_S\omega_S = \omega_A(N_S + N_R)$$

TABLE 6-1

Sun Gear	Arm	Ring Gear	$\dfrac{\omega_{OUT}}{\omega_{IN}}$	Speed Reduction or Increase
Input	Output	Fixed	$\dfrac{N_S}{N_S + N_R}$	Reduction
Output	Input	Fixed	$\dfrac{N_S + N_R}{N_S}$	Increase
Fixed	Output	Input	$\dfrac{N_R}{N_S + N_R}$	Reduction
Fixed	Input	Output	$\dfrac{N_S + N_R}{N_R}$	Increase

To establish all the input-output arrangements possible, there are three arrangements:

a) Arm fixed, $\omega_A = 0$. This converts the planetary train to an ordinary gear train and need not be further discussed.

b) Ring gear fixed, $\omega_R = 0$. Then

$$N_S \omega_S = \omega_A (N_S + N_R)$$

and
$$\frac{\omega_A}{\omega_S} = \frac{N_S}{N_S + N_R}$$

c) Sun gear fixed, $\omega_S = 0$. Then

$$N_R \omega_R = \omega_A (N_S + N_R)$$

and
$$\frac{\omega_A}{\omega_R} = \frac{N_R}{N_S + N_R}$$

The table shows that a speed reduction results if the output is taken from the arm. For a speed increase the arm must be the input. The greatest speed changes are produced by using a stationary ring gear.

Example. Determine suitable tooth numbers for a planetary train of the type shown in Fig. 6-7. The sun gear input rotates at 675 rpm and the output at the planetary carrier must be 150 rpm.

Solution. For this case

$$\frac{\omega_{OUT}}{\omega_{IN}} = \frac{N_S}{N_S + N_R} = \frac{150}{675}$$

Then N_S is proportional to 150 and $N_S + N_R$ is proportional to 675. If $N_S = 150$ then $N_R = 525$, and $N_R / N_S = 3.5$.

For a tooth number ratio of 3.5, suppose $N_S = 30$ is selected. Then $N_R = 3.5 \times 30 = 105$. For the number of teeth on the pinion the epicyclic tooth relationship must be used:

$$N_S + 2N_P = N_R.$$

With the tooth numbers already selected, $2N_P = 75$. Therefore, N_S must be some number such as 28, 24, or 32 if the planet gear is to have a whole number of teeth.

A method often used to analyze planetary trains sets up a tabulation of separate motions. It can best be explained by working the previous example.

In this method, the planetary train is first locked so that the gears cannot rotate, and the arm and the whole train are rotated one positive turn. But in the example the ring gear is actually fixed and ought not to have turned. A correction is made by unlocking the train and rotating the ring gear one negative turn to return it to zero rotation. When this is done, the planet and sun gears must also rotate the numbers of turns dictated by their tooth numbers. All this tabulates as follows:

	Arm	Ring	Planet	Sun
Lock train and rotate arm	+1	+1	+1	+1
Rotate ring one negative turn	0	−1	$-\dfrac{144}{24}$	$+\dfrac{144}{96}$
Total rotations	+1	0	−5	+2.5

The sun gear to arm speed ratio is 2.5:1, and the arm rotates $2/5 \times 1200$ rpm or 480 rpm in the same direction as the sun gear.

6.5.   THE BEVEL GEAR DIFFERENTIAL.

The basic bevel gear differential is shown in Fig. 6-8. This is actually a planetary train if the input side gear is taken as the sun gear, the two spider gears mounted on the spider as planet gears, and the fixed left-hand side gear as a ring gear. For this differential, using an input rpm of 1000,

$$\frac{\omega_R - \omega_A}{\omega_S - \omega_A} = -\frac{N_S}{N_R}$$

$$\frac{0 - \omega_A}{1000 - \omega_A} = -\frac{50}{50}$$

$$\omega_A = +500 \text{ rpm}.$$

The output shaft rotates in the same sense as the input shaft.

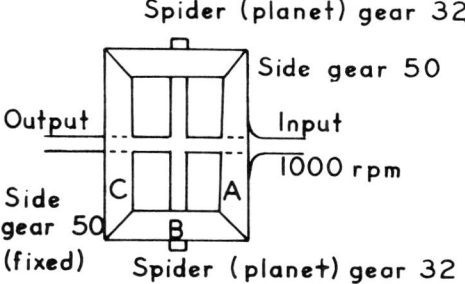

FIG. 6-8 BASIC DIFFERENTIAL.

To understand the operation of the differential, consider instantaneous velocities with a fixed side gear C as in Fig. 6-8. Let a be the velocity of the point of contact of gears A and B, and let bc designate the instant center for gear B. Then the velocity of the center of gear B is $a/2$, since C is fixed. Then

$$\omega_A = a/R_A \text{ and } \omega_{ARM} = a/2R_A. \quad \text{(See Fig. 6-9.)}$$

In the same way, if gear A is fixed and gears B and C rotate, with b being the velocity of the point of contact, then the velocity of the center of gear C

$$= \omega_C = b/R_C \quad \text{and} \quad \omega_{ARM} = b/2R_C \quad (R_C = R_A)$$

If both A and C rotate, then depending on direction of rotation, velocities $a/2$ and $b/2$ will either add or substract.

$$\frac{a/2 + b/2}{R} = \omega_{ARM} = \frac{\omega_A + \omega_C}{2}$$

so that $\omega_A + \omega_C = 2\omega_{ARM}$. This is the equation of the differential.

If ω_A and ω_C are equal in magnitude and sense, then $\omega_A = \omega_C = \omega_{ARM}$. Examination of the differential mechanism itself shows that this must be true since A and C will carry the spider arm around with them. If ω_A and ω_B are equal in magnitude but opposite in direction then $\omega_{ARM} = 0$.

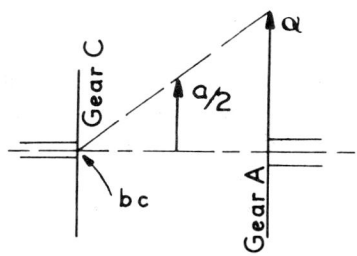

FIG. 6-9 VELOCITIES IN A DIFFERENTIAL.

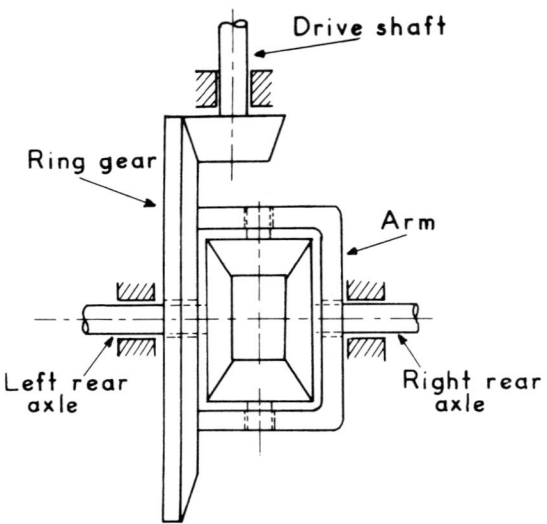

FIG. 6-10   AUTOMOBILE DIFFERENTIAL.

The differential used in the rear ends of automobiles is illustrated in Fig. 6-10. When the vehicle moves along a road in a straight line, the whole differential rotates as a unit, rotating both rear axles at the same speed: $\omega_A = \omega_C = \omega_{ARM}$. When making a turn, the outside rear wheel increases its speed, and the inside wheel must decrease in speed by the same amount if the tires are not to slip. From the equation of the differential it may be seen that for a given value of ω_{ARM}, if ω_A is increased (or decreased) then ω_C must decrease (or increase) by the same amount. Again, if one rear wheel slips on ice while the opposite wheel is on pavement, the wheel on the pavement does not turn, while the wheel on ice rotates at twice the speed of the arm.

In torque relationships, too, the differential has some unusual characteristics. The spider torque is always twice the torque of either side gear, and the torques of the two side gears are always equal:

$$\frac{T_{ARM}}{2} = T_A = T_C$$

Thus for the case where one wheel of a vehicle is on ice and the other wheel on pavement, the torque at the wheel on pavement is limited to the torque delivered by the wheel on ice, which is small and insufficient to move the vehicle.

These speed and torque relationships are independent of the tooth numbers in the differential.

6.6. SPEED CHANGE GEAR BOXES.

The usual power sources for driving machines are AC (alternating current) induction motors or internal combustion engines operated at a constant speed. At the output end of the machine the speed requirements frequently can be satisfied by a few speed settings. A restricted number of machines, such as welding positioners, extrusion machines, and some sheet rolling mills require infinitely variable speed control. If a limited range of output speeds is satisfactory, then the constant-speed input shaft can be connected to the output shaft through the medium of a speed change gear box. Familiar applications for such gear boxes are tractors, road vehicles, and machine tools such as lathes and milling machines. If the prime mover is an internal combustion engine, then the gear box must provide a reverse gear, but for an electric motor it is more economical to reverse the rotation of the motor.

A common type of gear box is shown in Fig. 6-11. This gear box offers six speeds. Gears 1 and 2, on the input shaft, and gears 5, 6, and 7, on the intermediate shaft, are fixed gears. The other gears are cluster gears which slide axially on splined shafts to engage the fixed gears as required. Two speeds of the intermediate shaft are available by engaging either gear 3 or gear 4 with the input shaft; with the three gears of the output shaft six speeds are available.

It is usual to provide output speeds arranged in geometrical progression. Suppose that the input shaft of Fig. 6-11 is to be driven at 1500 rpm, that the slowest speed desired is 200 rpm, and that the highest output speed is the speed of the input shaft, 1500 rpm. Then

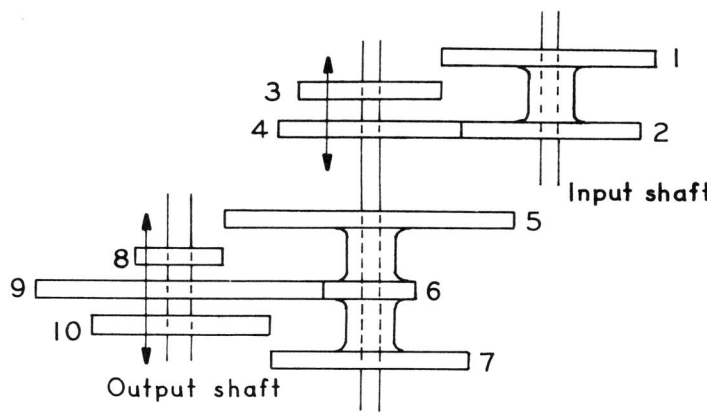

FIG. 6-11 SIX-SPEED GEAR BOX.

the overall speed ratio is $1500/200 = 7.5$. The ratio between any two speeds in a six-speed gear box will be

$$\sqrt[5]{75} = 1.496.$$

The fifth root is used and not the sixth root because there are five intervals between six speeds. Therefore the selected speeds are

$$
\begin{aligned}
&= 200 \text{ rpm} \\
200 \times 1.496 &= 299.2 \text{ (presumably taken as 300 rpm)} \\
299.2 \times 1.496 &= 447.6 \\
447.6 \times 1.496 &= 669.6 \\
669.6 \times 1.496 &= 1001.7 \\
1001.7 \times 1.496 &= 1498.6 \text{ (1500 rpm)}
\end{aligned}
$$

Since exact speeds are not required, the speeds would probably be taken as 200, 300, 450, 670, 1000, and 1500 rpm if tooth ratios allow these figures.

The gear box of Fig. 6-11 has a two-component train followed by a three-component train. The two-component train should drive the intermediate shaft at two speeds, one being the high speed of 1500 rpm. This high speed can be geared down to provide speeds of 1002 and 670 rpm. The other intermediate shaft speed could be 448 rpm, from which the low speeds of 448, 299, and 200 rpm can be obtained.

The number of teeth in the gears will be governed by the space available for the gear box and the required diametral pitch. Coarser teeth are required for higher torques. The following tooth numbers might be suitable for each gear:

Gear No.	Teeth
1	65
2	30
3	65
4	100
5	65
6	40
7	52
8	65
9	90
10	78

Gears 2 and 4 will rotate the intermediate shaft at 450 rpm.

The actual speeds are these:

Gear train	Ratio	Speed
$(1-3) \times (5-8)$	$65/65 \times 65/65$	1500
$(1-3) \times (6-9)$	$65/65 \times 40/90$	667
$(1-3) \times (7-10)$	$65/65 \times 52/78$	1000
$(2-4) \times (5-8)$	$30/100 \times 65/65$	450
$(2-4) \times (6-9)$	$30/100 \times 40/90$	200
$(2-4) \times (7-10)$	$30/100 \times 52/78$	300

6.8. FREQUENCY OF TOOTH CONTACT.

Gear teeth are subject to wear. Here we consider a matter of distributing this wear uniformly among the teeth. Consider two mating gears with the same number of teeth, say 21/21. For this pair, the same tooth on one gear will always drive the same tooth on the other gear. Wear will be greater because tooth contact will always occur in the same area of the tooth. If the two mating gears have 21 and 42 teeth, then in every second revolution the same pair of teeth will come into contact. This is better, but does not sufficiently distribute the wear. If however 21 teeth engage 43 teeth, any given pair of teeth will mate only every 21×43 or 903 revolutions. This makes for a much better distribution of wear. Therefore, an extra tooth or hunting tooth is recommended, or often required. The extra tooth usually makes it impossible to obtain the exact speed required, but this is not commonly a matter of any consequence.

The addition of the hunting tooth is an extra though minor complication in the design of gear trains. For the sake of simplicity of presentation, this matter was omitted in the examples in this chapter, but must not be ignored in practice.

PROBLEMS

1. Design compound spur gear trains for the following speed reductions. Reversal of rotation will be executed at the electric motor driving the train, and therefore direction of rotation can be ignored.
    a) Speed reduction 1:20. Minimum number of teeth 20 and maximum 120. Odd numbers of teeth above 25 are more expensive.
    b) Speed to be reduced from 1500 to 120 rpm. Minimum and maximum number of teeth allowable are 18 and 54. Odd numbers of teeth above 25 are more expensive.
    c) Speed reduction of 1:11. Minimum and maximum numbers of teeth allowable are 18 and 54. As before, odd numbers of teeth over 25 are more expensive.

2. Design a compound spur gear train for a speed reduction of 1:20 using not less than 18 or more than 96 teeth. The output gear must rotate in the direction opposite to the input gear.

3. A pair of mating spur gears has a diametral pitch of 8. The driving gear rotates at 1800 rpm and the output speed of the driven gear must be as close to 700 rpm as possible, using a center distance as close to 4.500 in. as possible. Find the number of teeth in each gear, the output speed, and the center distance.

4. A pair of mating spur gears has a driver rotating at 900 rpm and a driven gear at 400 rpm. Center distance must be 3.500 in. closely. A diametral pitch in the range of 10 to 14 inclusive must be used. Decide the best diametral pitch, number of teeth, pitch diameters, and output speed.

5. A welding turntable is driven by the gear train shown in the figure, with input power supplied by an infinitely variable-speed electric motor. A fillet weld is to be deposited around the circumference of a 6.000 diameter steel tube rotated by the turntable, at a welding speed of 16 inches per minute. What must be the speed of the driving motor?

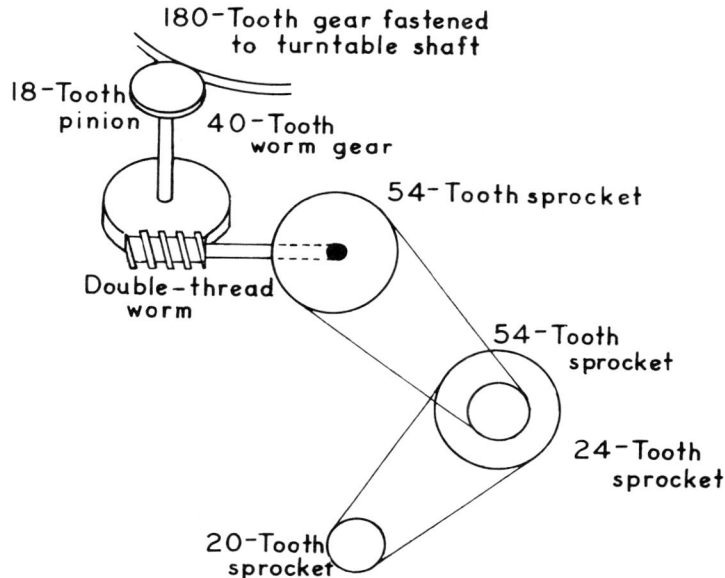

FIG. P6-5 GEAR TRAIN FOR A WELDING TURNTABLE.

6. Design a compound spur gear train for a speed reduction of 8.25:1 using tooth numbers from 16 to 96 inclusive.

7. Design a spur gear train for a speed ratio within the range of 0.2943 to 0.2945. Minimum and maximum tooth numbers are 16 and 96.

8. A belt drive with a center distance of 9.75 in. must be replaced by a gear drive. The existing pulleys have pitch diameters of 3.6 in. on the input shaft and 7.7 in. on the output shaft. In attempting replacement with gears, the center distance cannot be altered, nor can the direction of rotation of the shafts—in a belt drive the shafts rotate in the same direction.

Attempt to design a suitable spur gear drive, using a diametral pitch of 12 or coarser teeth. If using a 20° pressure angle, a minimum of 18 teeth is desirable. A gear drive that does not exceed the belt drive in total length on centerline is desirable. Sketch the drive.

9. Design reverted gear trains to the following specifications. State number of teeth in all gears and give center distance. Use a uniform diametral pitch throughout each drive.
   a) Minimum and maximum number of teeth are 20 and 100; center distance is 3.20 in. closely; speed reduction is 1:10.
   b) Speed reduction is 2:25; minimum and maximum number of teeth are 18 and 120; diametral pitch is 12.

10. Design reverted gear trains for the following speed reductions. Allowable minimum and maximum numbers of teeth are 15 and 96.
   a) 1:12. Diametral pitch is 6 for the first reduction and 5 for the second.
   b) 1:11. Diametral pitch is 12 for the first reduction and 10 for the second.
   c) 1:18. Diametral pitch is 8 for the first reduction and 5 for the second.

11. Determine the planet carrier shaft speed if the sun gear of Fig. 6-7 is the input, the ring gear is fixed, and the planet carrier is the output, given the following information: $N_s = 96, N_p = 24, N_R = 144, \omega_s = 1200$ rpm clockwise.

12. If in Fig. 6-7 the input shaft at the sun gear rotates at 1800 rpm clockwise, what is the output and direction of rotation at the planet carrier? The sun gear has 60 teeth, and the planet gear 15 teeth.

13. By fixing sun gear or ring gear and by using either as input or output, what four possible speed ratios are obtainable from the planetary gear train of Problem 12?

14. The following tooth numbers apply to a planetary train similar to that of Fig. 6-7: sun gear 20, ring gear 100. Determine the four possible speed ratios obtainable by fixing either sun or ring gear and by using either as output or input.

15. Design an epicyclic gear train of the three-gear type including ring gear to provide the following speed reductions:
   a) 9:1
   b) 11:1

State number of teeth in gears and which members are input and output.

16. Design a planetary gear train of the type shown in Fig. 6-7 to provide a speed ratio increase of 6:1. Give tooth numbers and state which members are input and output.

17. In the figure, the cluster gears of 48 and 32 teeth are keyed to the input shaft. The 48-tooth gear in the ordinary train drives the planetary arm, while the 64-tooth gear in the ordinary train drives the first sun gear of 35 teeth. Determine the output speed of the train if the input speed is 100 rpm.

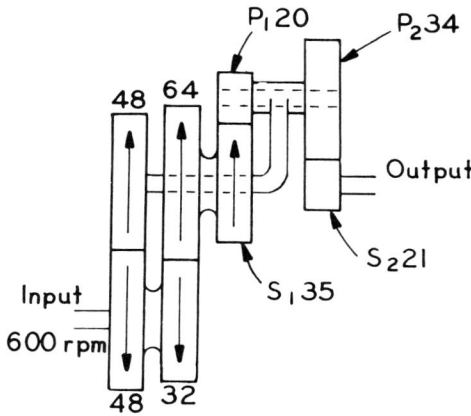

FIG. P6-17 REVERTED ORDINARY AND PLANETARY TRAIN.

18. a) Explain why an epicyclic gear train is so named.
     b) The epicyclic mechanism of the figure consists of a fixed ring gear with a pitch diameter of 4.000 in. and a planet gear with a pitch diameter of 1.000 in. on a carrier arm. There is no sun gear. Determine the path traced by a point *P'* on the planet gear at a radial distance of 0.197 in. below the axis of rotation of the planet gear when the arm is in the vertical position shown.

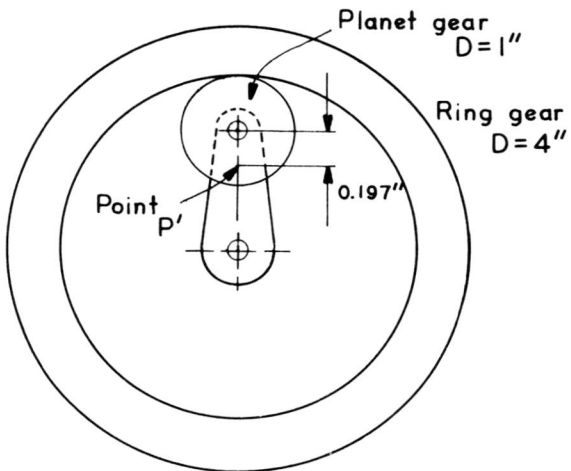

FIG. P6-18   EPICYCLOID MECHANISM.

19. In the planetary train shown in the figure the arm "floats" between input and output shafts, and gears *B* and *D* are fastened to the same short shaft on the arm. Determine the

speed ratio between input and output shaft and the rotation of the output shaft with respect to the input shaft.

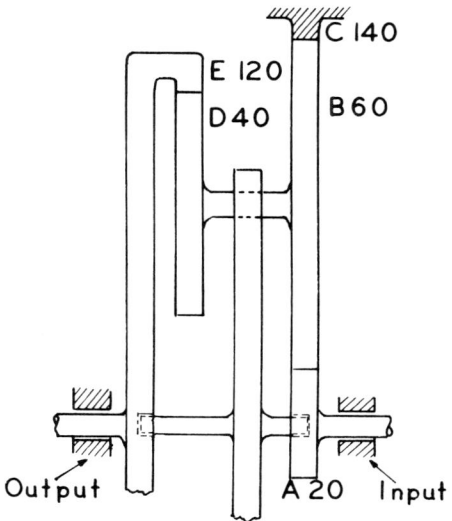

FIG. P6-19 A REVERTED PLANETARY TRAIN.

20. For the epicyclic train of the figure, what is the output/input speed ratio, using either shaft as input?

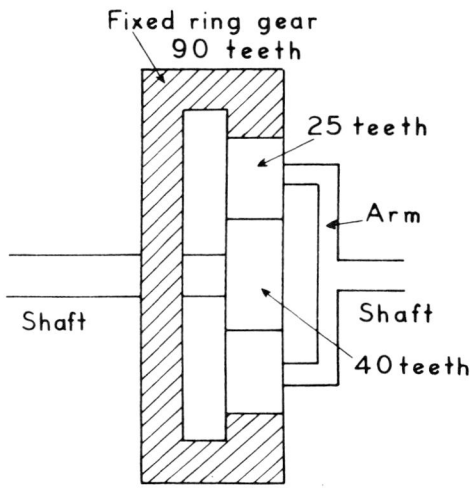

FIG. P6-20 PLANETARY TRAIN.

21. Design a basic planetary train of the type shown in Fig. 6-7 to produce a speed reduction of 5, minimum number of teeth 18.

22. Design a basic planetary train of the type shown in Fig. 6-7 to produce a speed increase of 3.50. Use a diametral pitch of 24 and a ring gear pitch diameter approximating 3.0 in.

23. Design a planetary train for a speed reduction of 100/160 using a stationary sun gear.

24. For a reverted planetary train of the design shown in Fig. 6-6 the following tooth numbers are used:

    gear A—90 teeth   gear C—92 teeth
    B—91 teeth   D—91 teeth.

    Determine the ratio of output to input speed for the train.

25. In the reverted planetary train of the figure, the ring gear is fixed and the output is taken from the planetary arm. The 80- and 25-tooth gears are fastened to a shaft in the arm and rotate together. Determine the output/input speed ratio and the relative direction of rotation.

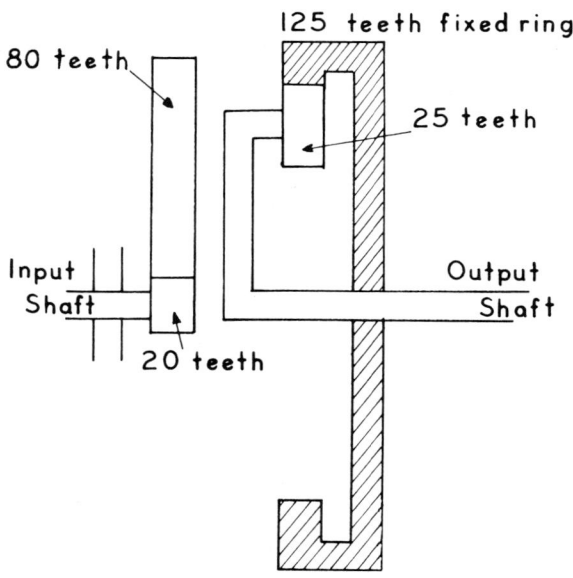

FIG. P6-25   REVERTED PLANETARY TRAIN.

26. In the gear train of the figure, gears *B* and *C* are both fastened to the input shaft; gear *A* turns in its own bearing and meshes both with gear *B* and the ring gear *E*; the planetary pinion *D* rotates on the arm and meshes both with gear *C* and the ring gear *E*. Both the ring gear and the arm rotate independently about the same axis of rotation, the arm being connected to the output shaft. Determine the output/input speed ratio and the sense of rotation of the output shaft relative to the input shaft.

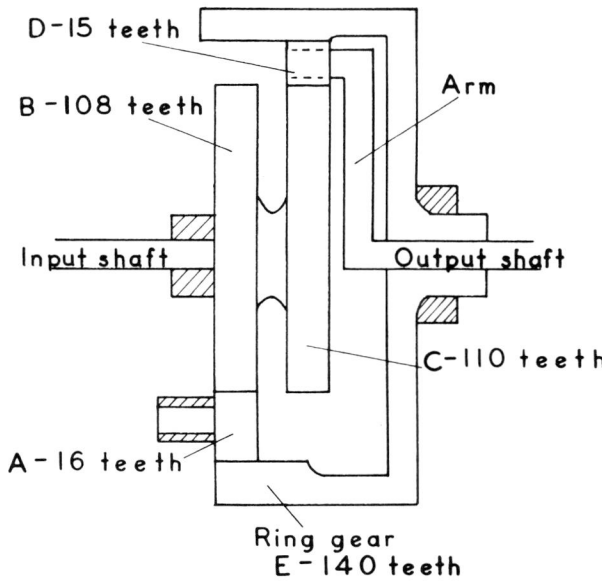

FIG. P6-26   PLANETARY TRAIN.

27. In the epicyclic train of the figure, the ring gear *D* is fixed. Planet gears *B* and *C* are fastened to the same shaft, rotating at the end of the arm. The output is taken from the arm. Determine the output/input speed ratio and the directions of rotation.

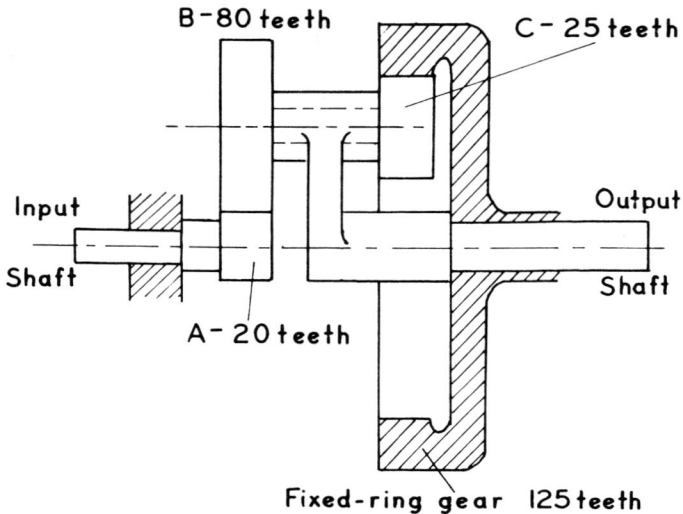

FIG. P6-27   REVERTED PLANETARY TRAIN.

28. In the epicyclic train of the figure, tooth numbers are

gear A—30 teeth   gear C—15 teeth
   B—75 teeth   D—120 teeth.

The ring gear D is fixed and the planetary arm is fastened to the input shaft. If the input shaft rotates clockwise at 1000 rpm, find the speed of the output shaft.

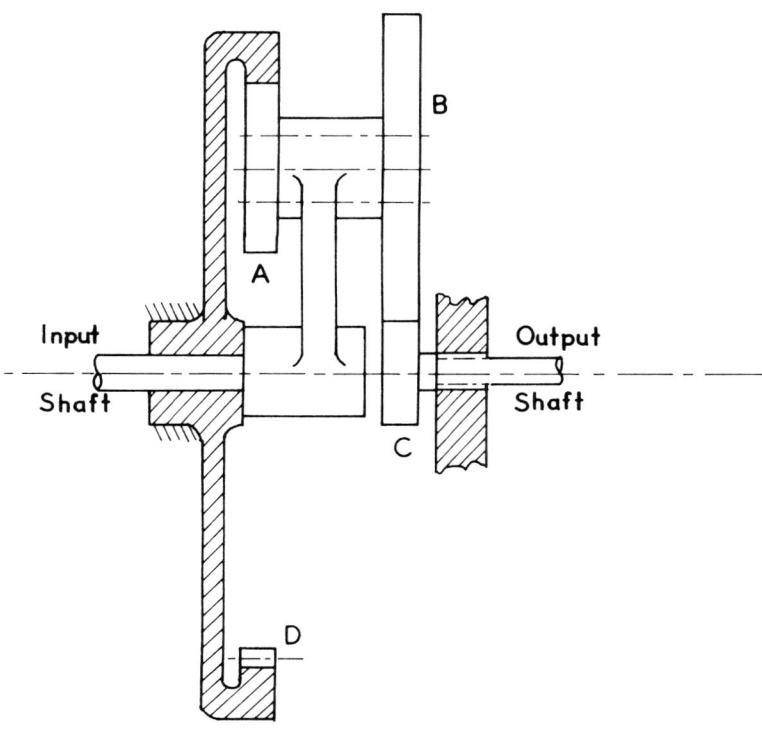

FIG. P6-28   REVERTED TRAIN.

29. The reverted epicyclic train of the figure uses bevel gears. The ring gear *C* is fixed. Determine the output/input speed ratio.

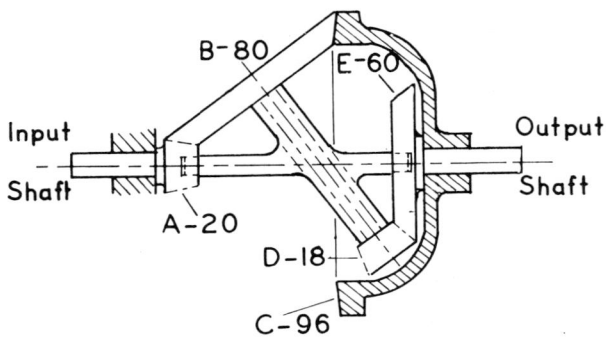

FIG. P6-29   BEVEL GEAR EPICYCLIC TRAIN.

30. In the gear train of the figure, the sun gear *D* is fixed; gears *C* and *E* are fastened to a shaft rotating in the end of the planetary arm. Determine the output/input speed ratio.

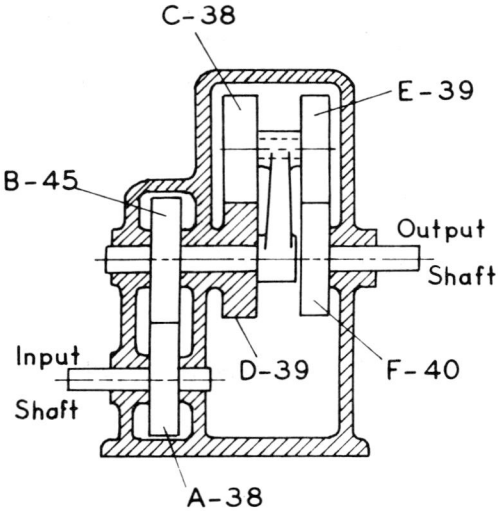

FIG. P6-30   ORDINARY AND EPICYCLIC TRAIN.

31. An automobile makes a right-hand turn on a curve with a radius of 40 ft to the center of the rear-axle differential. The rear wheels of the vehicle are 5 ft apart on centers, and tire diameter is 28 in. The bevel gear on the drive shaft (see Fig. 6-10) has 18 teeth and the ring gear driving the spider has 54 teeth. Find the following for a drive shaft speed of 900 rpm:
    a) The rpm of the right rear wheel
    b) The rpm of the left rear wheel
    c) The speed of the automobile in mph

32. In the differential of the figure, all gears have 75 teeth. The arm rotates at 100 rpm; the gear C is fixed. Find the rpm for gear A.

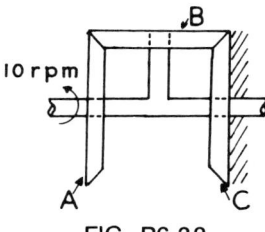

FIG. P6-32

33. In the bevel gear train of the figure, gears B and C are fastened to the same shaft on the arm. Determine the output/input speed ratio. Gear D is fixed.

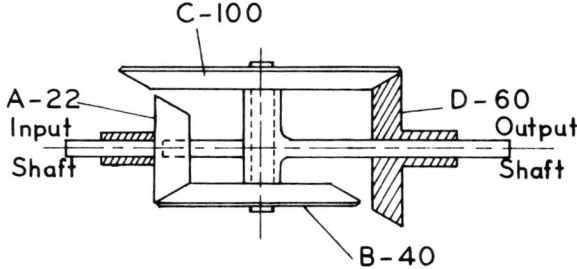

FIG. P6-33   BEVEL GEAR PLAENTARY TRAIN.

34. The gear train of the figure is driven by an input shaft rotating at 1800 rpm counterclockwise. The output is taken from the 22 tooth gear. Due to belt creep, each of the belt drives loses 3% from its ideal speed ratio. Make an accurate determination of the speed of the output shaft, indicating its direction of rotation.

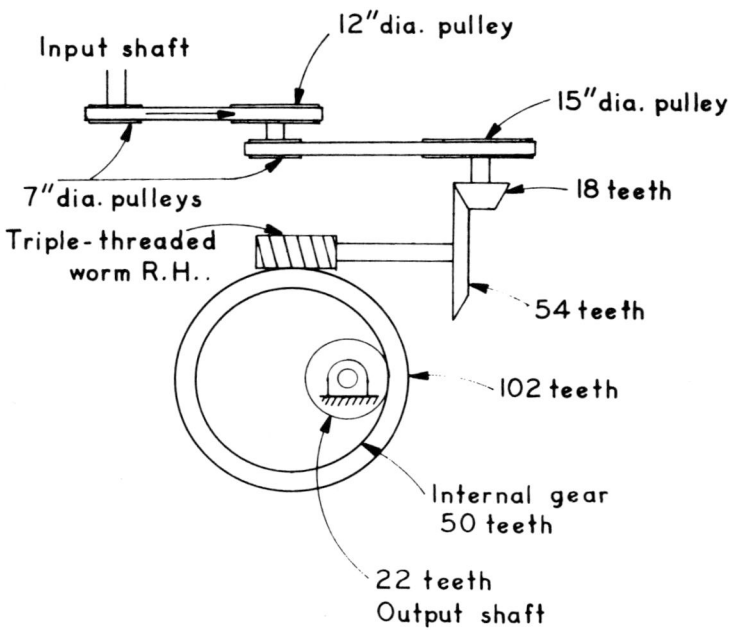

FIG. P6-34   BELT AND GEAR DRIVE.

35. The gearbox of the figure provides nine speed ratios. All gears on the countershaft are fastened to that shaft. The group of cluster gears *ABC* slide on splines so that *A* can engage *E* or *C* can engage *G*, while *B* is shown engaging *F*. Similarly, the cluster gears *IJK* are spline-mounted so that *I* can engage *D*, *J* can engage *F*, or *K* can engage *H*. Determine all possible speed ratios.

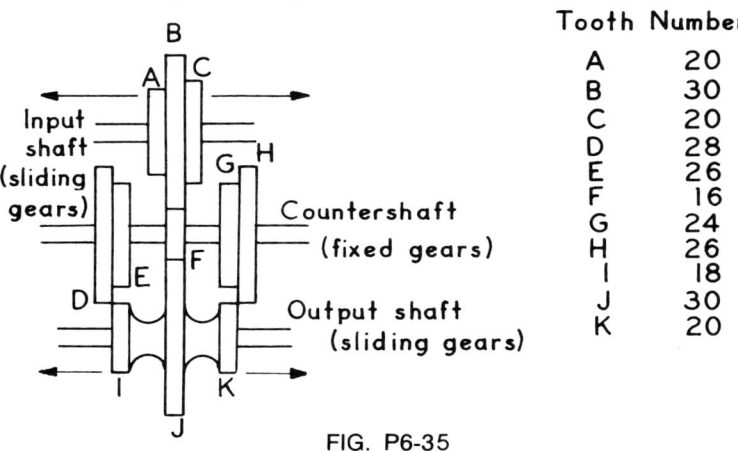

Tooth Numbers

A	20
B	30
C	20
D	28
E	26
F	16
G	24
H	26
I	18
J	30
K	20

FIG. P6-35

36. Design a speed change gear train for an input speed of 1800 rpm and output speed of 300, 1000, and 1350 rpm using parallel shafts. The output shaft must rotate in the same sense as the input shaft. Select from the following stock gear sizes only: every tooth number from 12 to 30; every even number from 30 to 100; every fifth number from 100 to 125; every tenth number from 130 to 160.

37. The figure shows a four-speed transmission using spur gears of diametral pitch 7 and a center distance of 4.286 in.

Gear Train	Speed Ratio, Output/Input
1–2–5–8	0.156
1–2–4–7	0.324
1–2–3–6	0.592
straight through	1.000

Find the tooth numbers for all gears.

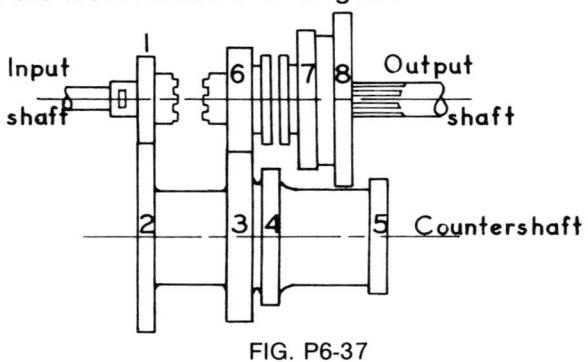

FIG. P6-37

38. A transmission using the same arrangement of gears as for Problems 37 has a center distance of 7.000 in. and a diametral pitch of 6. Speed ratios are the following:

Gear Train	Speed Ratio, Output/Input
1–2–5–8	0.250
1–2–4–7	0.308
1–2–3–6	0.500
straight through	1.000

Find the tooth numbers for all gears.

39. Design a six-speed gear box for a machine tool with speeds ranging in geometrical progression from 100 to 1500 rpm and with an input shaft speed of 900 rpm. Diametral pitch is 8.

40. Design a six-speed gear box for a milling machine with an input shaft speed of 450 rpm and a speed range of 60 to 930 rpm, approximately. Diametral pitch is 7.

41. Design a six-speed gearbox for an input speed of 830 rpm and output speeds from 205 to 615 rpm, using a geometric progression of output speeds.

chapter seven

Automation Devices

7.1. AUTOMATION AND KINEMATICS

"Automation" means automatic material handling, parts handling, or tool handling. There are three types:

1. Processing of bulk liquids and solid materials in the type of automation called *process control*, as in the automatic control of a blast furnace, an oil refinery, or chemical processes.

2. *Numerical control*, which controls automatic workheads and workhandling devices by means of punched tape, punched cards, magnetic tape, or direct connection to a computer.

3. *Fixed automation*, which is the handling and processing of parts in a fixed, predetermined, and repetitive cycle. Spark plugs, for example, must be manufactured in quantities of millions of units, and in these quantities must be automatically manufactured.

Only fixed automation requires the techniques of kinematic analysis and the hardware of kinematics such as cams, gears, slider

231

cranks, and four-bar linkages. The techniques of designing an automatic work station are not a proper subject for this book, but we can examine those automation devices that have kinematic interest. In addition to the design of a required displacement, velocity, and acceleration, automation devices impose one more requirement: movement must be synchronized with the movements of other devices belonging to the automatic machine. For example, an escapement device which releases a screw to an automatic screwdriver must do so only when the screwdriver is ready to receive the next screw.

An automatic parts feeding system consists of the following components:

1. Some kind of feeding device that constantly feeds a supply of parts, such as screws.

FIG. 7-1 IN-LINE AUTOMATIC ASSEMBLY OF SPRAY NOZZLES USING ELECTRIC AND PNEUMATIC CONTROLS. (*DIXON AUTOMATIC TOOL, INC.*)

2. An orienting device that removes all the fed parts that do not have the required orientation. For example, screws must be supplied head up to an automatic screwdriver, and all screws that are supplied end up or lying on their sides must be removed from the feeding system.

3. An escapement, which is a mechanism that escapes parts only when the machine is ready to receive them.

The parts will be fed to various types of workheads, such as automatic screw drivers, riveters, staplers, presses, or brazing heads. The complete automatic work station will be a coordination of many of these subsystems, such as the one shown in Fig. 7-1. Often the system is built around a circular indexing table, the assembly being completed by indexing through each of the positions of the indexing table.

We begin with a discussion of two of the more important automatic feeding devices.

7.2. CENTERBOARD HOPPER.

The centerboard hopper of Fig. 7-2 is of simple construction, and its motion is less complex than that of the vibratory bowl feeder discussed below. The centerboard hopper will not feed parts that tend to tangle, such as springs (springs can be fed from a vibratory bowl feeder), nor certain flat shapes such as washers, which also can be fed from a vibratory bowl feeder. The centerboard hopper has a hopper into which a load of parts is dumped, and an oscillating

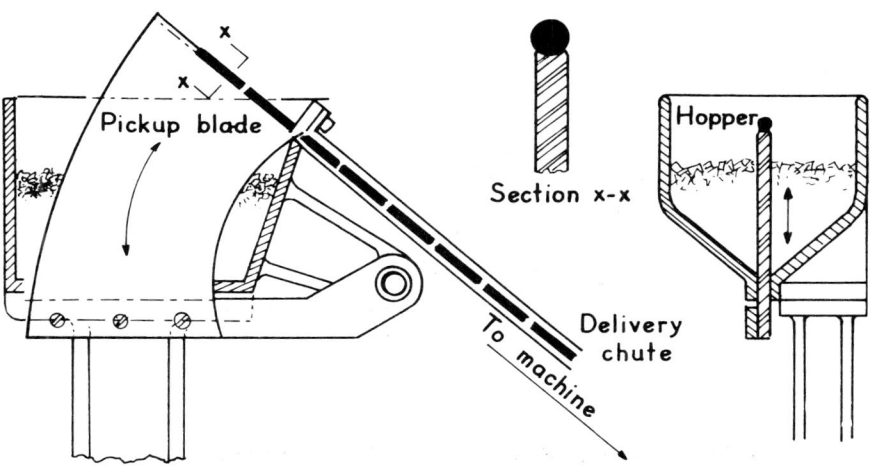

FIG. 7-2 CENTERBOARD HOPPER.

pickup blade. The top surface of the blade has a track shaped to hold the parts picked up when the blade rises through the hopper. For cylindrical parts the track would have a concave cross section. The blade catches a few parts, and when it is in its highest position, it is aligned with a chute. The caught parts slide down the track into the chute for delivery to the workhead.

Some type of drive, cam or other, must operate the oscillating blade. At its lowest point, the blade track must be aligned with the bottom of the hopper. It must rise from this position to catch a few parts in its track, and at its highest position it must dwell for a sufficient time period to allow the caught parts to slide from the track. Deceleration to the highest position must be controlled, otherwise the parts will leave the track. The emptied track must be returned rapidly to its bottom position.

If the angle θ_{max} (Fig. 7-3) is increased, then the dwell time to allow parts to slide from the track can be decreased. But simultaneously a longer period of time is required to complete the upward motion of the blade.

The possibility of a part leaving the track as the blade decelerates to discharge position is greatest at the end of the track farthest from the pivot point.

For the reaction between the track and the part to be zero the following relation must hold (Fig. 7-4)

$$ma\left(R - \frac{l}{2}\right) + mg\cos\theta_{max} = 0$$

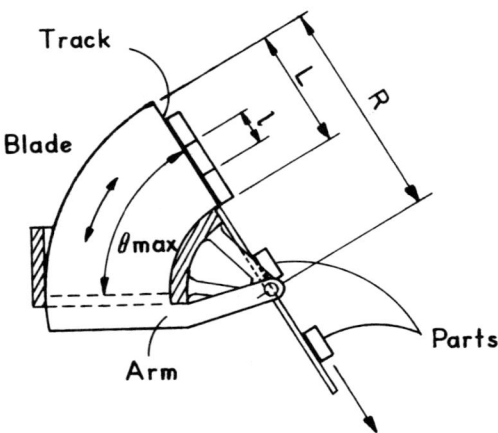

FIG. 7-3   CENTERBOARD SIGNIFICANT DIMENSIONS.

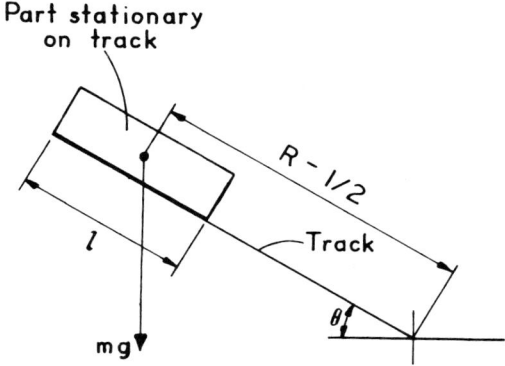

FIG. 7-4 PART LYING ON THE TRACK.

where
$\quad m$ = mass of the part = w/g
$\quad R$ = maximum radius of the track
$\quad \theta_{max}$ = maximum angle of the track to the horizontal
$\quad \alpha$ = angular acceleration of the track

For convenience assume that $R - l/2 = R$. Since l is much smaller than R, this assumption introduces no great error. Then

$$R\alpha = g\cos\theta_{max}$$

and the maximum possible angular deceleration must be

$$\alpha = \frac{g\cos\theta_{max}}{R}$$

The mechanism driving the blade must give it a motion approximating this:

a constant acceleration $\qquad \dfrac{g\cos\theta_{max}}{R}$

followed by

a constant deceleration $\qquad \dfrac{g\cos\theta_{max}}{R}$

Using this maximum value for acceleration and deceleration, the time required to raise the blade to angle θ_{max} can be determined. The first half of the angle θ_{max} is traversed during an acceleration from zero angular velocity and the second half during a deceleration to zero angular velocity. Use the general equation

$$\theta = \frac{1}{2}\alpha t^2$$

Substitute $\frac{1}{2}\theta_{max}$ for θ and t_1 for the time of acceleration or of deceleration, to give:

$$\frac{\theta_{max}}{2} = \frac{1}{2}\alpha t_1^2$$

$$\theta_{max} = \alpha t_1^2$$

But

$$\alpha = \frac{g\cos\theta_{max}}{R}$$

$$\theta_{max} = \frac{g\cos\theta_{max}}{R}t_1^2$$

$$t_1^2 = \frac{R\theta_{max}}{g\cos\theta_{max}}$$

The total time to move the blade through angle θ_{max} is twice t_1. Let this total time be T_1, where $T_1 = 2t_1$. Then

$$T_1^2 = \frac{4R\theta_{max}}{g\cos\theta_{max}}$$

At angle θ_{max} there must be a dwell just sufficient to allow the caught parts to slide down the track to the discharge chute. The longest sliding time is that for sliding the full length of the track. Let $L = $ length of the track. The forces acting on a part sliding down the track are shown in Fig. 7-5. Taking force components parallel to the track

$$ma = mg\sin\theta_{max} - \mu mg\cos\theta_{max}$$

where $a = $ linear acceleration of the part down the track and μ is the coefficient of friction between part and track. To find the minimum dwell time use the basic equation

$$s = \frac{1}{2}at^2 \qquad \text{or} \qquad t^2 = \frac{2s}{a}$$

and substitute the track length L for s. For t, T_2 will be substituted, T_2 being the minimum dwell time to allow parts to slide the length of the track.

$$T_2^2 = \frac{2L}{g(\sin\theta_{max} - \mu\cos\theta_{max})}$$

The time for the return of the blade to the bottom of the hopper must be the shortest time period possible. If this return time is ignored, then the total period of the oscillation

$$= T = T_1 + T_2 = \left(\frac{4R\theta_{max}}{g\cos\theta_{max}}\right)^{1/2} + \left(\frac{2L}{g\sin\theta_{max} - \mu\cos\theta_{max}}\right)^{1/2}$$

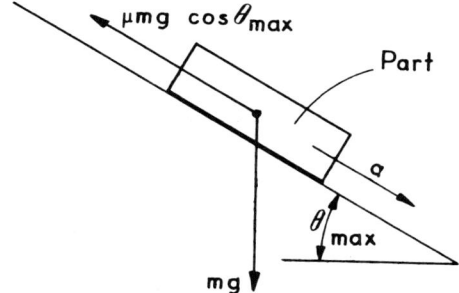

FIG. 7-5 FORCES ACTING ON A PART SLIDING DOWN
THE TRACK.

When this equation for cycle time is known, what angle θ_{max}
would provide the highest rate of feeding of parts? A mathematical
attack on this problem would be a formidable one. However, some
reasonable assumptions can be made:

1. That $R/L = 2.0$

2. That θ_{max} must be about 45°

3. That a reasonable figure for μ is 0.3

Using $R = 10$, $L = 5$ in., $\theta_{max} = 45°$ or 0.785 radians, and $\mu = 0.3$

$$T = \left(\frac{4 \times 10 \times 0.785}{32.2 \times 0.707}\right)^{1/2} + \left(\frac{2 \times s}{32.2 \times 0.707 - 0.3 \times 0.707}\right)^{1/2}$$
$$= 1.176 + 0.666 = 1.84 \text{ sec}$$

If the parts are 1 in. long, then ideally the track length can
collect 5 parts if it is 5 in. long. This will not happen. A reasonable
figure is 2 parts collected per cycle. This gives a feed rate of slightly
better than 1 part per second. A longer track will give a somewhat
higher feed rate. Quick-return cams or cranks operate the centerboard at about 40 cycles per minute.

7.3. THE VIBRATORY BOWL FEEDER.

An important application of simple harmonic motion is the
vibratory bowl feeder (Fig. 7-6). This is a hopper feeder for small
parts. The bowl diameter should be not less than 6 times the length
of the part to be fed. A helical track climbs the wall of the bowl at a
small helix angle of about 4°. The bowl is supported on three sets of

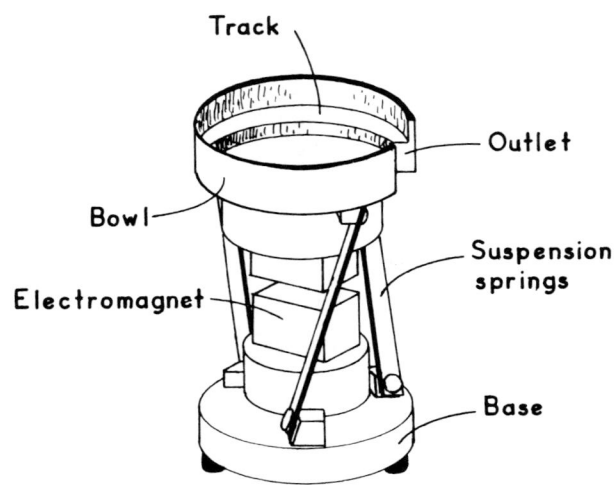

FIG. 7-6 VIBRATORY BOWL FEEDER. THREE SUCH
BOWLS MAY BE SEEN IN THE IN-LINE
ASSEMBLY MACHINE OF FIG. 7-1.

inclined flat springs as shown in Fig. 7-6. The bowl is vibrated by an
electromagnet mounted in the base of the feeder. Because of the
angular mounting of the supporting springs, both a vertical and a
torsional (horizontal) vibration is imparted to the bowl. Since 60-
cycle alternating current is usually supplied to the electromagnet,
the vibratory cycle is a case of simple harmonic motion.

To understand how a part can travel up the bowl track, consider that when the bowl begins a cycle of vibration the part lying on
the track is forced to follow the forward (torsional) motion of the
track because of friction with the track. Without friction parts could
not be fed. The part acquires sufficient inertia so that when the bowl
retracts on the negative half-wave of vibration, the part continues in
the forward direction, being finally stopped by friction. Thus the
part makes one forward hop per cycle of vibration. For very small
amplitudes of vibration insufficient inertia is developed and the part
remains stationary on the track (Fig. 7-7). Control of the amplitude
of vibration is obtained by controlling the amount of electric current
to the windings of the electromagnet.

For simple harmonic motion force is proportional to displacement:

Force = mass × acceleration = $-kx$, where × = displacement

The frequency of vibration=f, usually 60 cycles/sec, and $\omega = 2\pi f$

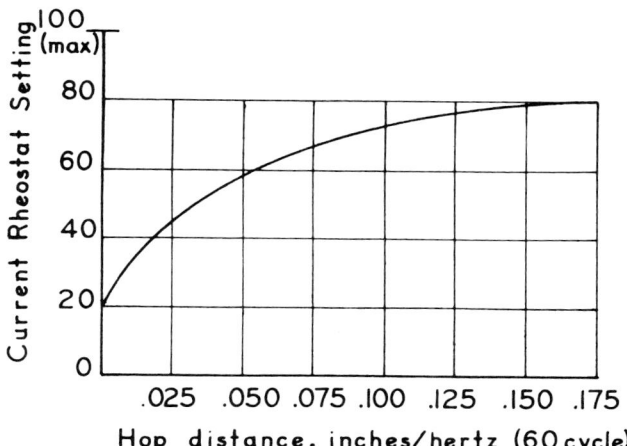

FIG. 7-7 LENGTH OF HOP PER HERTZ OF 60 HERTZ
POWER MADE BY PARTS IN A VIBRATORY
BOWL FEEDER.

rad/sec   ($=377$ if $f=60$). Also maximum acceleration$= -\omega^2 A$,
where $A =$ maximum amplitude.

Fig. 7-8 is a free body diagram of a part on the track at
maximum amplitude A. The track angle is designated θ and the
vibration angle, combining both the horizontal and vertical components of vibration, is designated α. For sliding up the track to occur
the following conditions must hold:

$$m A \omega^2 \cos\alpha > mg \sin\theta + F$$

But
$$F = \mu N = \mu(mg\cos\theta - mA\omega^2\sin\alpha)$$

Then
$$m A \omega^2 \cos\alpha > mg \sin\theta + \mu(mg\cos\theta - mA\omega^2\sin\alpha)$$

$$A\omega\cos\alpha > g\sin\theta + \mu(g\cos\theta - A\omega^2\sin\alpha)$$

$$A\omega^2(\cos\alpha + \mu\sin\alpha) > g\sin\theta + \mu g\cos\theta$$

Therefore
$$\frac{A\omega^2}{g} > \frac{\sin\theta + \mu\cos\theta}{\cos\alpha + \mu\sin\alpha}$$

But since θ is about 4°, $\cos\theta$ is closely 1, so that

$$\frac{A\omega^2}{g} > \frac{\sin\theta + \mu}{\cos\alpha + \mu\sin\alpha}$$

For a track angle of 4° $\sin\theta$ will be 0.07, and μ will be in the range of
0.2 to 0.3 for a metal track, or in a higher range if the track is
rubber-lined. ω^2/g for 60-cycle power is 4444.

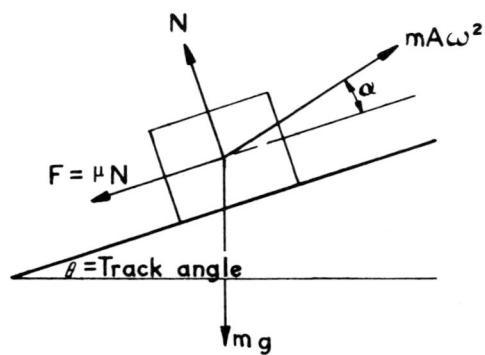

FIG. 7-8   FREE-BODY DIAGRAM OF A PART IN A VIBRA-
TORY BOWL FEEDER.

By similar analysis it can be shown that for backward sliding
down the track

$$\frac{A\omega^2}{g} > \frac{\mu\cos\theta - \sin\theta}{\cos\alpha - \mu\sin\alpha}$$

The above analysis indicates that conveying velocity is independent of the weight of the part. The factors that influence this
velocity are these:

1. Track angle. The influence of track angle is slight for the
   range of angles in use.

2. Vibration angle.

3. Coefficient of friction. Higher coefficients improve the conveying rate.

4. Track acceleration.

5. Operating frequency.

If the vibration amplitude is sufficiently large then the part
will overcome gravitational pull and leave the track or hop. For this
to occur, the normal force N must become zero, that is

$$N = mg\cos\theta - mA\omega^2\sin\alpha = 0$$

Then
$$g\cos\theta = A\omega^2\sin\alpha$$

$$\frac{A\omega^2}{g} = \frac{\cos\theta}{\sin\alpha}$$

The hopping action gives a higher feed rate.

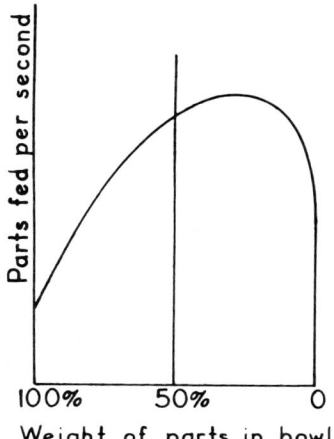

FIG. 7-9   CHARACTERISTIC CURVE OF FEED RATE AS
THE VIBRATING BOWL EMPTIES.

The vibratory bowl feeder does not feed at a constant rate. The power input is constant, and therefore as the bowl gradually feeds out parts there is less mass to accelerate, causing the amplitude of vibration gradually to increase with a resulting increase in the rate at which parts are fed. A typical graph of feed rate as the bowl empties is shown in Fig. 7-9. The drop at the end of the feed rate curve occurs because the bowl becomes nearly empty and is starved of parts to feed.

Besides the centerboard hopper and the vibratory bowl feeder there are other feeding devices. The others have minimal interest for the subject of kinematics and are not discussed here.

7.4.   ORIENTING DEVICES.

There is a great variety of standard and special orienting devices for presenting parts in proper orientation.

For feeding from a vibratory bowl feeder, orientors may be in-bowl or out-of-bowl. Out-of-bowl tooling is fitted to the trough or chute connecting the bowl feeder to the workhead. In-bowl tooling uses the general principle of diverting misoriented parts back into the bowl to be recycled.

The design of orientors is basically governed by the unique characteristics of the part to be fed, such as its shape, position of center of gravity, pattern of holes, or other characteristics. Certain types of orientors however have some general application. Three of these, all in-bowl devices, must be discussed. These are the wiper blade, the pressure break, and the slot. They are best understood in terms of feeding screws or bolts.

Suppose machine screws are fed up the track of a vibratory bowl feeder. The range of possible orientations of such parts will comprise screws

a) Standing on their heads

b) Lying on their sides across the track

c) Lying on their sides with head leading

d) Lying on their sides wiith thread leading.

Screws cannot be fed in the desired orientation, that is thread down or head up. Further, the screws will not necessarily feed in single file, but may be doubled or in a tangled condition with one screw lying on another. The flow of screws must be oriented so that all are in single file and with head up (thread down). Three devices will produce the required orientation: the wiper blades, the pressure break, and the slot (Fig. 7-10).

The train of screws first encounters the first wiper blade. This blade sweeps back into the bowl any screws standing vertically on their heads. The height of the wiper blade is so positioned as to pass all screws lying on their sides.

The next item of tooling is the pressure break. This closes off a portion of the width of the track, leaving only a narrow width for the screws to pass through. The pressure break will direct back into the bowl any tangled screws and those oriented at right angles to the track. Past the pressure break only those screws remain on the track that are oriented longitudinally to the track, with either head or tail

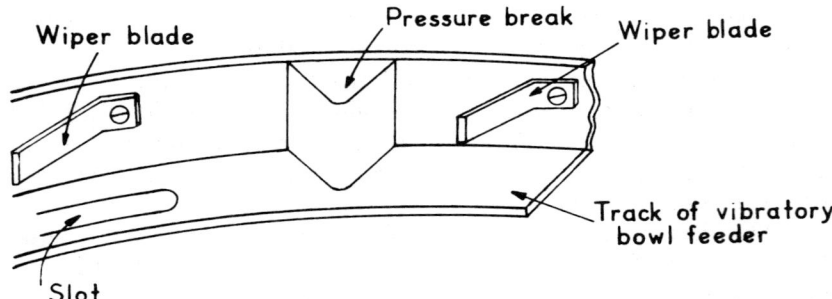

FIG. 7-10   IN-BOWL ORIENTING DEVICES.

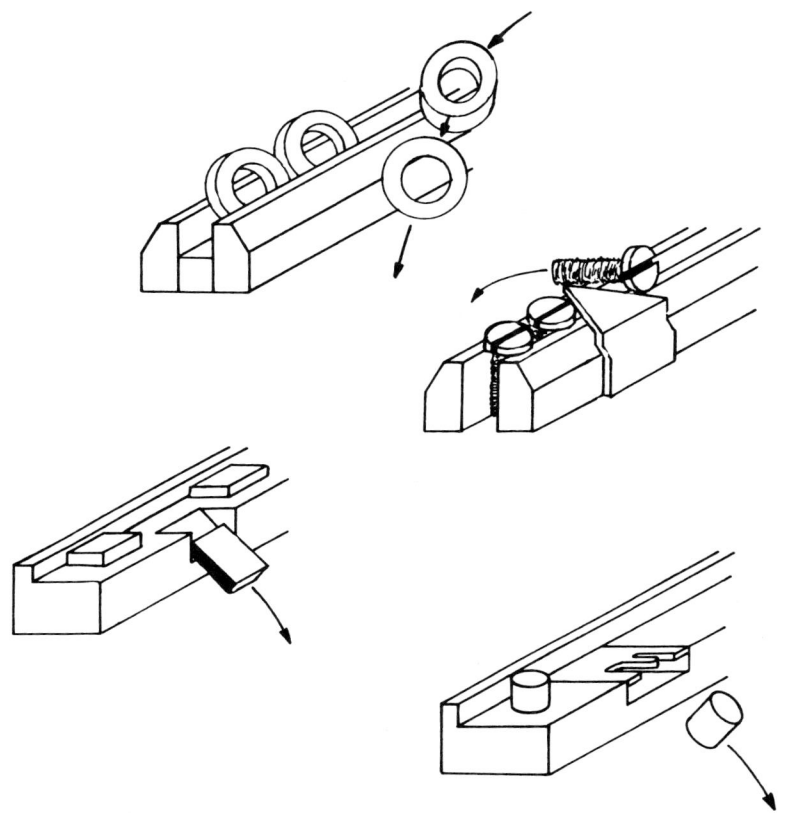

FIG. 7-11   A VARIETY OF ORIENTING DEVICES.

leading. With either orientation, almost all screws will drop their tails into the slot. There are, however, always a few recalcitrant ones. These can be rejected into the bowl for recycling by the second wiper blade, which is positioned just high enough to pass the screw head.

Screws and bolts are easy to orient; other kinds of parts may be more difficult. Coil springs can tangle, but the vibration of the bowl tends to untangle them. Washers are a simple shape that should require no orientation; they will feed flat and are normally required in this position. But one washer may be lying on another, and the two can then wedge under a wiper blade to block the flow of washers. An air jet instead of a wiper blade can be used to blow off a washer piled on another washer.

Other orienting devices are shown in Fig. 7-11.

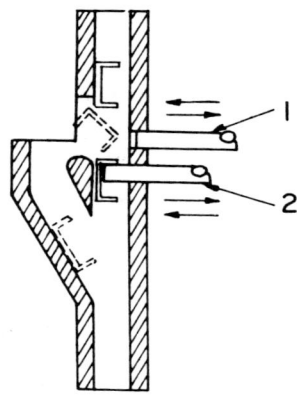

FIG. 7-12   ORIENTATION OF U-SHAPED PARTS.

The orientation of a channel-shaped part is shown in Fig. 7-12. The parts drop down the feed tube with the open side facing either left or right. The two pins marked 1 and 2 reciprocate horizontally. Every part falls to sit on pin 2. Pin 1 then advances into the tube. If the open side of the channel is facing to the right, then the motion of pin 1 has no effect, and as pin 2 withdraws, the part falls through. If the open side of the channel is facing to the left, then pin 2 kicks over the part as indicated by the dotted lines. Thus all parts are delivered with the open side facing the same way. A variant of the above device is the orienter for cup-shaped parts, Fig. 7-13.

The design of suitable orienting devices requires ingenuity and imagination applied to an examination of the characteristics of the part.

7.5.   ESCAPEMENTS.

An escapement must release a part (or several parts), such as a screw, from a bank of these parts when a mechanism such as a power screwdriver is ready to receive the part. Many escapements take the general form of the one of Fig. 7-14. Two pins or arms reciprocate in sequence. The rear arm moves down to hold back the bank of parts in the delivery chute while the forward arm lifts to release a part to the workhead. The forward arm then moves down, and the rear arm lifts to allow the bank of parts to shift forward. The arms must be shaped to suit the length and contour of the parts (Fig. 7-15). Such escapements may be operated by a solenoid, a small air cylinder, or a mechanical arm. Only a short stroke of about $\frac{1}{2}$ in. is needed in the actuator. Other variants of this device are shown in Fig. 7-16.

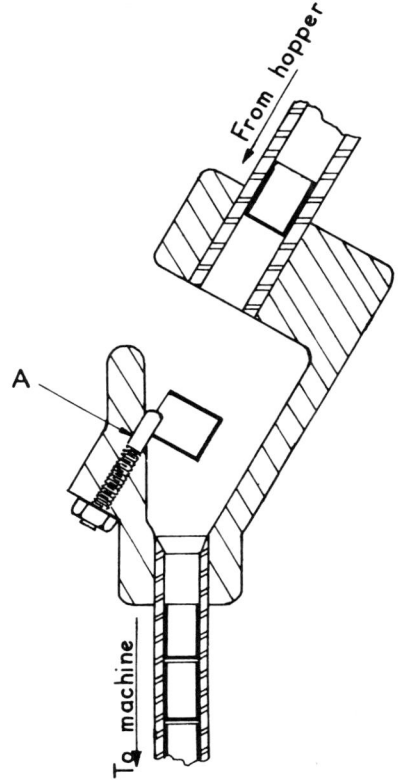

FIG. 7-13   DEVICE FOR ORIENTING CUP-SHAPED PARTS
OPEN-END UP. A PART WITH THE REQUIRED
ORIENTATION PASSES THROUGH; A PART UP-
SIDE DOWN IS CAUGHT ON PIN *A* AND
TURNED AROUND.

The gate escapement of Fig. 7-17 supplies two work stations,
but cannot supply both simultaneously.

Sometimes the escapement can be incorporated into the
workhead. In Fig. 7-18 the parts feed from the delivery chute
directly to the workhead, the leading part hanging in the jaws ready
for the punch. When the punch descends to drive the part it blocks
off the bank of parts. When the punch retracts another part slides
into the jaws. A variation of this type of escapement is shown for the
automatic screwdriver of Fig. 7-19.

FIG. 7-14   ESCAPEMENT OPERATED BY A PNEUMATIC CYLINDER.

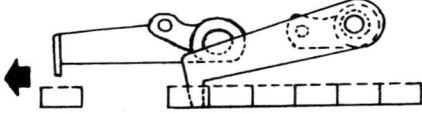

FIG. 7-15   ESCAPING CUP-SHAPED PARTS WITH THE OPEN END UP.

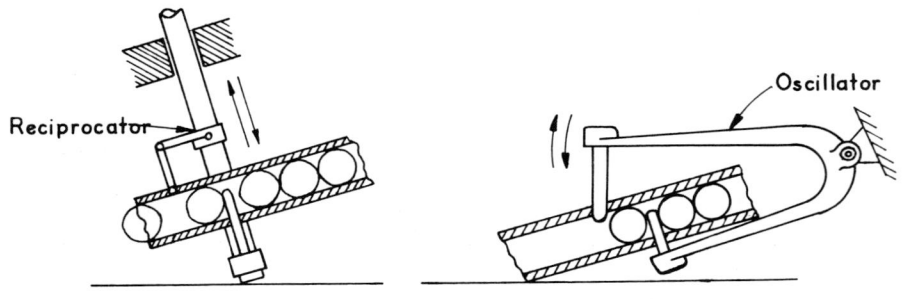

FIG. 7-16   VARIATIONS OF THE ESCAPEMENT DEVICE OF FIG. 7-11.

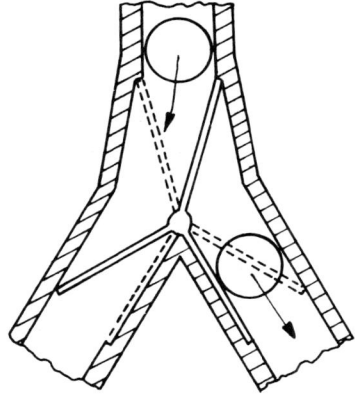

FIG. 7-17   GATE ESCAPEMENT FOR SUPPLYING TWO
WORKSTATIONS.

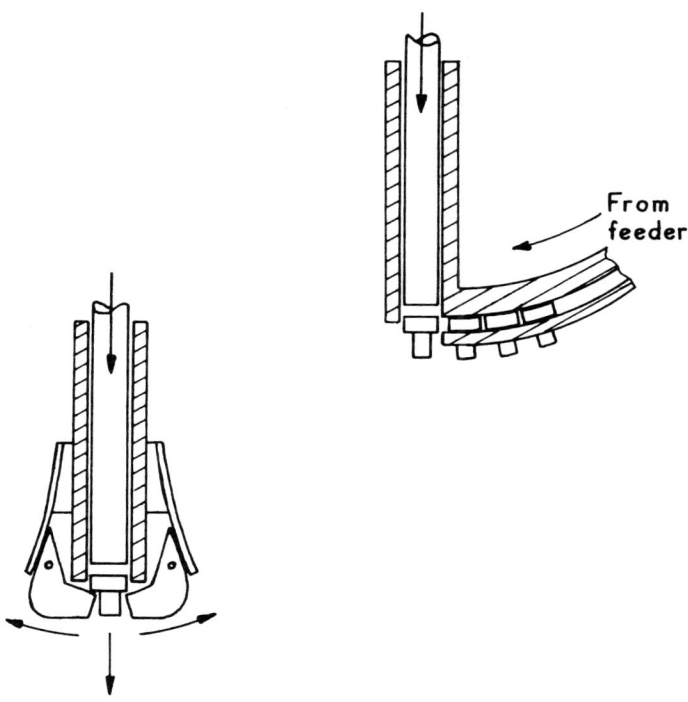

FIG. 7-18   ESCAPEMENT INCORPORATED IN THE
WORKHEAD OF A PUNCH.

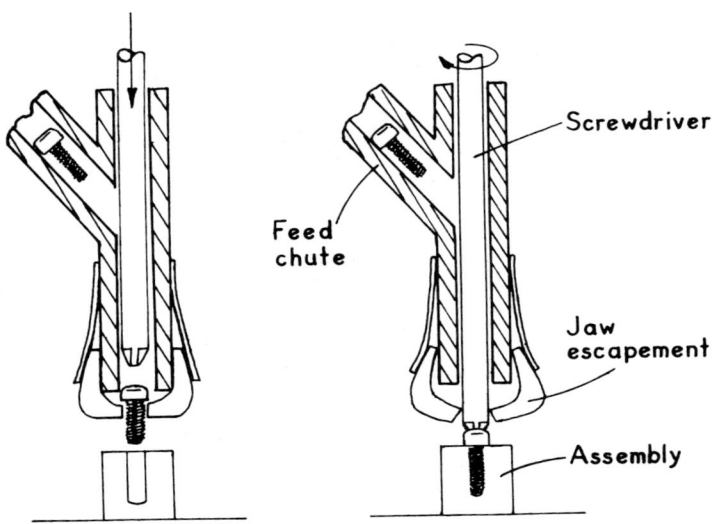

FIG. 7-19   A VARIATION OF FIG. 7-15, IN AN AUTOMATIC
SCREWDRIVER.

7.6.   PARTS PLACING DEVICES.

Sometimes a transfer mechanism is required to pick parts from
a track and place them in the work station. A variety of designs are
in use for this kind of transfer, some being quite complex. Fig. 7-20
shows a simple one. A mechanical, vacuum, or magnetic head on a
transfer arm picks up a part from the feed track, elevates it,
transfers it horizontally, and finally lowers the part into position.

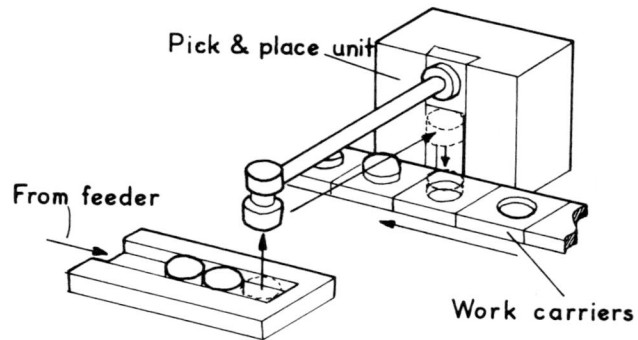

FIG. 7-20 "PICK AND PLACE" PARTS PLACING
MECHANISM.

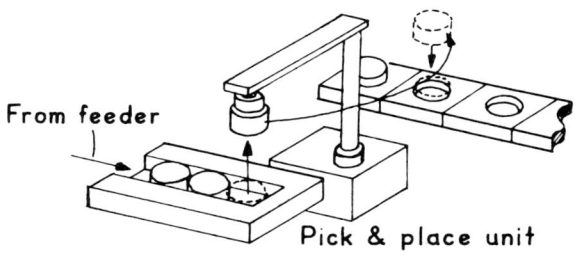

From feeder

Pick & place unit

FIG. 7-21   ROTATING PICK AND PLACE MECHANISM.

The head then returns along the same path to its initial pickup position. A variant of this device is shown in Fig. 7-21. Here the transfer arm rotates 180 degrees about a vertical axis. The column has axial movement also.

A more elaborate cam-operated electromagnetic device is shown in Fig. 7-22. The cam and mating gears rotate the arm and its electromagnet through 90°. The part is picked up by the electromagnet from the feed chute and elevated above the vertical guide chute. When the electromagnet is switched off, the part is dropped into the chute.

The placing mechanism of Fig. 7-23 has both a vertical and a horizontal slide, each operated by its own cam. Both cams are mounted on the same shaft. The pickup head may be a vacuum cup or an electromagnet. Both cam follower arms have elongated slots

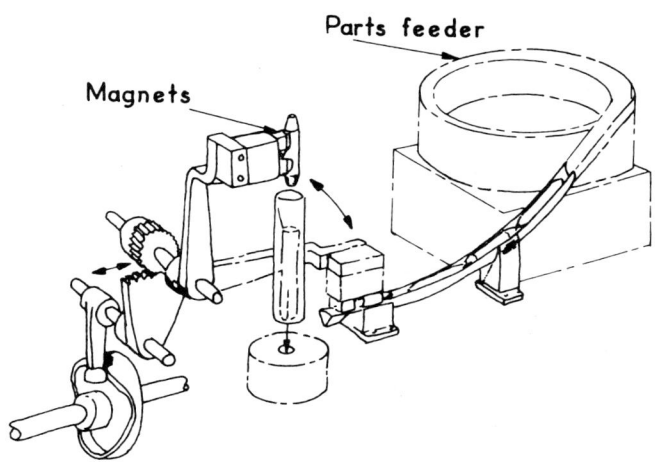

Parts feeder

Magnets

FIG. 7-22   ELECTROMAGNETIC PICKUP DEVICE.

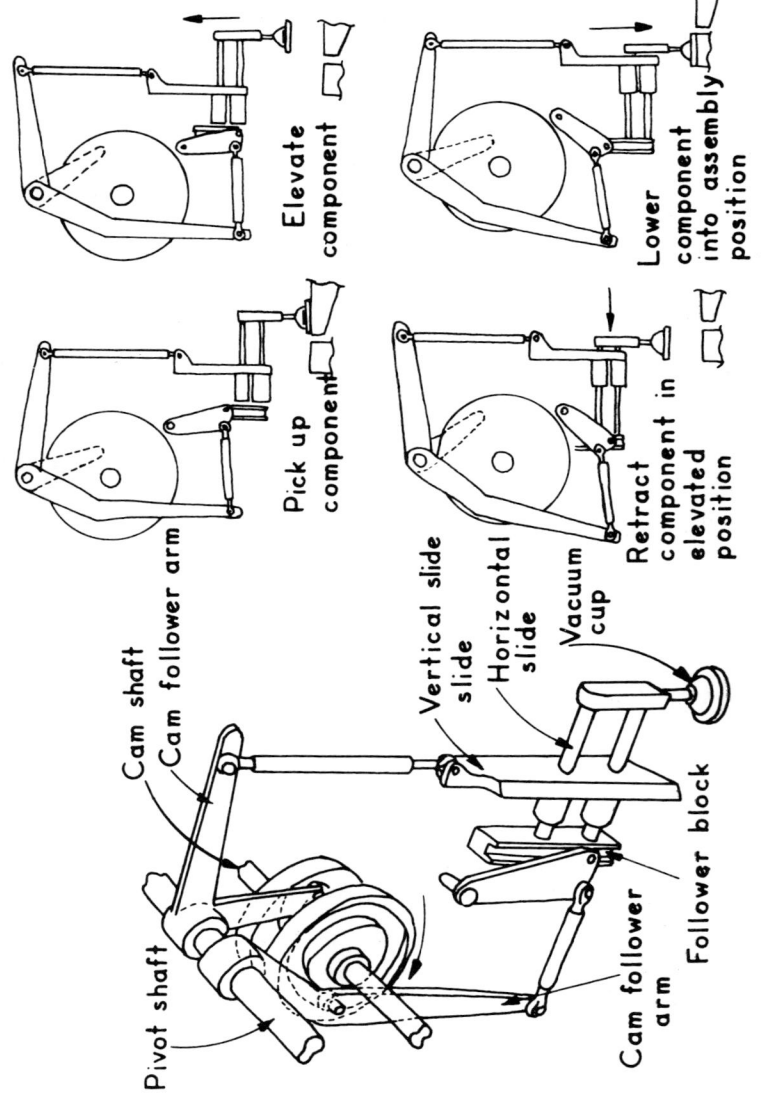

Elevate component

Lower component into assembly position

Pick up component

Retract component in elevated position

Cam shaft
Cam follower arm

Pivot shaft

Vertical slide slide

Horizontal slide

Vacuum cup

Cam follower arm

Follower block

FIG. 7-23 PLACING MECHANISM WITH BOTH VERTICAL AND
HORIZONTAL MOTION BY MEANS OF TWO CAMS.

for accurate adjustment of the mechanisms. The sequence of movements is the following:

1. Forward movement of the horizontal slide and downward movement of the vertical slide so that the pickup head can contact the part

2. Upward movement of the vertical slide to raise the part

3. Backward movement of the horizontal slide

4. Lowering of the part to the assembly position

The pickup mechanism of Fig. 7-24 has the same sequence of motions but a somewhat different design. It too has a cam for each movement, both cams mounted on the same shaft.

The mechanism of Fig. 7-25 picks a part from the bottom of a vertical stack of parts. There is an inner shaft within an outer sleeve shaft, each shaft being operated by its own cam as shown in the figure. The vacuum cup picks a part from the bottom of the stack of parts; both shafts move down in synchronism to lower the part by a distance greater than the radius of the part. Then the inner shaft dwells while the outer shaft continues to move down. The pinion connected to the outer shaft then must rotate on the rack mounted on the inner shaft, thus rotating the vacuum cup 180°. When rotation is completed, both shafts move down again in synchronism to place the part in position.

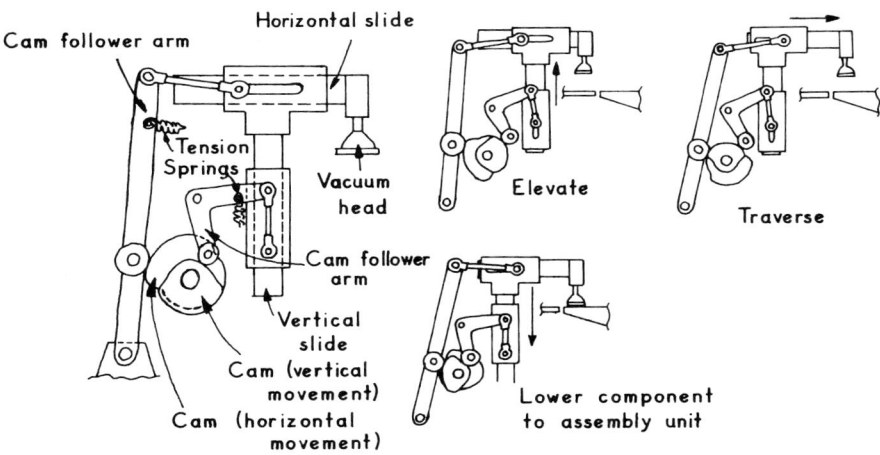

FIG.   7-24   ANOTHER   DOUBLE   CAM   PLACING MECHANISM FOR VERTICAL AND HORIZONTAL MOTION.

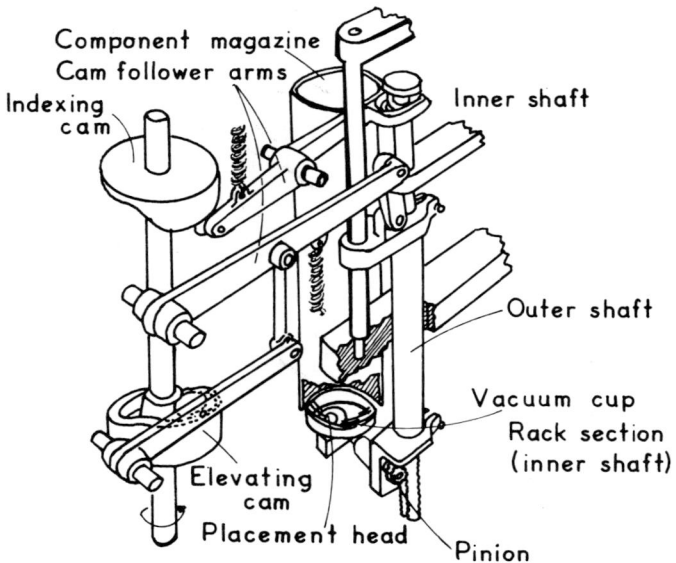

FIG. 7-25 MECHANISM FOR REMOVING A FLAT PART
FROM THE BOTTOM OF A STACK OF PARTS
IN A PARTS MAGAZINE.

Another method of removing a part from the bottom of a magazine of parts is given in Fig. 7-26. Again two cams are required. One cam raises and lowers the magazine of parts and the other cam extends and retracts the pickup head, which is a vacuum pickup. The sequence of operations is the following.

1. The parts magazine is lowered to the pickup head.

2. The pickup head captures the lowest part in the magazine.

3. The magazine is elevated out of the way of the pickup head.

4. The pickup head is moved forward and rotated 180° to deposit the part.

The rotation of the vacuum head is produced by a chain rack and mating sprocket, and two bevel gears shown.

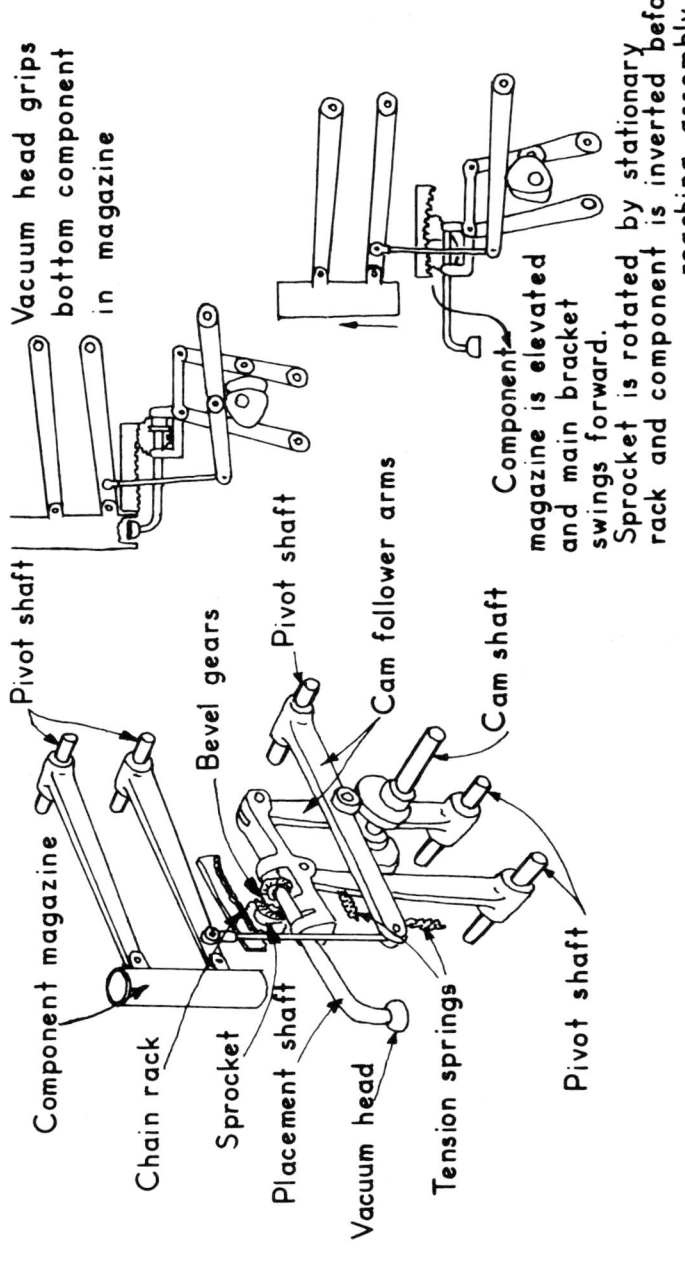

Vacuum head grips
bottom component
in magazine

Component
magazine is elevated
and main bracket
swings forward.
Sprocket is rotated by stationary
rack and component is inverted before
reaching assembly area.

Pivot shaft

Pivot shaft

Cam follower arms

Cam shaft

Component magazine

Bevel gears

Chain rack

Sprocket

Placement shaft

Vacuum head

Tension springs

Pivot shaft

FIG. 7-26   THIS MECHANISM PERFORMS THE FUNCTION OF THE ONE SHOWN IN FIG. 7–25 BUT HERE THE PARTS MAGAZINE IS NOT FIXED BUT MOVES VERTICALLY.

PROBLEMS

1. Using the free body diagram of the part on the track of a vibratory bowl feeder (Fig. 7-8), show that for the part to slide backward the following relationship must hold:

$$\frac{Au^2}{g} > \frac{\mu \cos\theta - \sin\theta}{\cos\alpha - \mu \sin\alpha}$$

   Since the vibration drives the part down instead of up, the direction of $mA\omega^2$ will be reversed.

2. Equate the expressions for the limiting conditions for forward and backward conveying in the vibratory bowl feeder and show that the limiting condition for forward conveying is given by

$$\tan\alpha > \frac{\sin\theta}{\mu^2}$$

3. Show that for the vibrated part to lose contact with the track or to hop, the following condition must be satisfied:

$$\frac{A\omega^2}{g} > \frac{\cos\theta}{\sin\alpha}$$

   Hint: Use $N = Mg\cos\theta - MA\omega^2\sin\alpha$.

4. Completely dimension a vibratory bowl track width, wiper blade, pressure break, and slot to orient 1/4 in. bolts 1 in. long with head up. Your solution is a trial solution only and would have to be tested in a vibratory bowl feeder.

5. Design an in-bowl orienting device (or sequence of devices) to orient the part shown in the figure big end down and to reject wrong orientations back into the bowl.

   Design Projects.

FIG. P7-5   PART TO BE ORIENTED.

1. Design a centerboard hopper for the feeding of tubular parts 15 mm in diameter by 30 mm long. The hopper itself must not be excessively wide and must be deep enough to hold sufficient parts. Design a suitable cam to operate the pickup blade at approximately 40 strokes per minute, preferably with a fast return stroke.

Estimate a realistic delivery rate of parts per minute for your design.

2. Design a mechanism similar to that of Fig. 7-22 for picking up and dropping steel cylinders into a delivery tube. The cylinders measure 12 mm in diameter by 35 mm long. Design a suitable cam and gearing, not necessarily the barrel cam of the figure, and select a suitable diametral pitch. Estimate a suitable dwell time and rpm for the camshaft, and avoid excessive acceleration in the cam design. The horizontal distance from the pickup position to the dropping position must be a minimum of 105 mm. Decide other dimensions as required.

3. Design a mechanism, of the type shown in Fig. 7-23, to pick up flat metal lids 30 mm in diameter. A horizontal movement of 80 mm is required between pickup and release positions. Any other dimensions are at the selection of the designer.

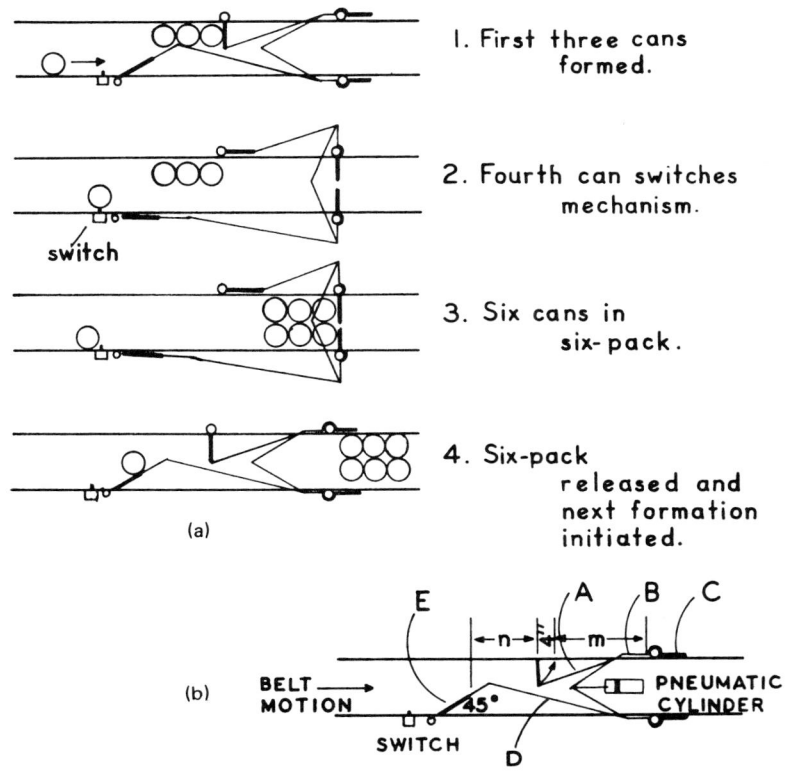

1. First three cans formed.

2. Fourth can switches mechanism.

3. Six cans in six-pack.

4. Six-pack released and next formation initiated.

(a)

(b)

FIG. P7-4 (A) SEQUENCE OF OPERATIONS FOR FORMING BEER CANS ON A BELT INTO A SIX-PACK FORMATION.
(B) DIMENSIONS OF SIX-PACK MECHANISM.

4. The accompanying series of figures (Fig. P7-4) shows the sequence of operations through which a linkage mechanism receives 6 beer cans and arranges them in a six-pack configuration. The mechanism is an ingenious combination of two four-bar linkages and two slider-crank mechanisms operated by a pneumatic cylinder. A sensing device and a counter count the beer cans on the conveyor belt as they enter the mechanism. The electrical counter and time delays activate the solenoid-operated pneumatic cylinder. This device was invented by J.B. Tokarski and K.M. Marshek at the University of Connecticut.

    Identify the four-bar linkages and the slider-crank mechanisms.

    Using the following basic dimensions, determine the lengths of all links and the required stroke of the pneumatic cylinder. For convenience the components are labelled by letters in Fig. P7-4(b). Use the following dimensions:
    1. Diameter of a beer can—3 in. (7.5 cm)
    2. Belt width—8.0 in. (20 cm)
    3. Elements B and C—4.0 in. long (10 cm)

        Your attention is drawn to the following requirements. (See Fig. P7-4b.)
    1. Distances *m* and *n* must accommodate 3 beer cans (9 in. or $22\frac{1}{2}$ cm).
    2. Determine the length of A from the position and dimensions shown in Fig. P 7–46.
    3. With *A, B,* and *C* known, the cylinder stroke can be determined.
    4. Diverter bar *E* crosses the belt at 45° to the belt.
    5. The length of *D* can be found by the solution of right triangles in its extreme positions.

        Check that your design dimensions are compatible in the two extreme positions of this mechanism.

Index

257